THE
COMPLETE
IDIOT'S
GUIDE® TO

Astronomy

Third Edition

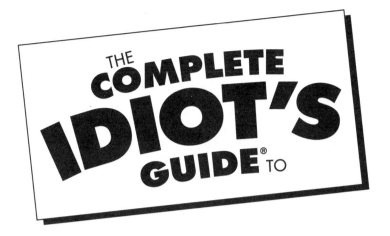

THE COMPLETE IDIOT'S GUIDE® TO

Astronomy

Third Edition

by Christopher De Pree, Ph.D.
and Alan Axelrod, Ph.D.

ALPHA

A member of Penguin Group (USA) Inc.

For Julia, Claire, and Madeleine (CGD)
For my stars, Anita and Ian (AA)

International Standard Book Number: 1-59257-219-7
Library of Congress Catalog Card Number: 2004101842

06   05   04   8   7   6   5   4   3   2

Interpretation of the printing code: The rightmost number of the first series of numbers is the year of the book's printing; the rightmost number of the second series of numbers is the number of the book's printing. For example, a printing code of 04-1 shows that the first printing occurred in 2004.

Printed in the United States of America

Note: This publication contains the opinions and ideas of its authors. It is intended to provide helpful and informative material on the subject matter covered. It is sold with the understanding that the authors and publisher are not engaged in rendering professional services in the book. If the reader requires personal assistance or advice, a competent professional should be consulted.

The authors and publisher specifically disclaim any responsibility for any liability, loss, or risk, personal or otherwise, which is incurred as a consequence, directly or indirectly, of the use and application of any of the contents of this book.

Most Alpha books are available at special quantity discounts for bulk purchases for sales promotions, premiums, fund-raising, or educational use. Special books, or book excerpts, can also be created to fit specific needs.

For details, write: Special Markets, Alpha Books, 375 Hudson Street, New York, NY 10014.

Publisher: *Marie Butler-Knight*
Product Manager: *Phil Kitchel*
Senior Managing Editor: *Jennifer Chisholm*
Senior Acquisitions Editor: *Mike Sanders*
Development Editor: *Jennifer Moore*
Production Editor: *Megan Douglass*
Copy Editor: *Molly Schaller*
Illustrator: *Richard King*
Cover/Book Designer: *Trina Wurst*
Indexer: *Julie Bess*
Layout/Proofreading: *Becky Harmon, Donna Martin*

Contents at a Glance

Appendixes

Contents

Appendixes

Foreword

I was going to be a marine biologist until my parents bought me a telescope when I was in the seventh grade. I took it outside and set it up in my backyard in suburban Fort Worth, Texas. The sky was clear and the stars were out. One bright star caught my attention. I pointed the small telescope at it to figure out why it was so brilliant. Pointing that little telescope took a bit of work, but I finally centered the bright dot in the finder scope and carefully looked through the main eyepiece. What I saw changed my life forever.

Instead of just a bright speck made brighter by the light gathering power of the telescope, what appeared was a small, bright, crescent-shaped object. I was floored. I had no idea what I was looking at. It looked kind of like the crescent Moon, but was much smaller and had no surface features. I ran inside to get the guidebook that came with the telescope and within a few minutes had figured out that I was looking at the planet Venus. I ran inside again, got my parents and brother and sister to come outside and showed them what I had discovered. I am not sure they were as impressed as I was. At least none of them became astronomers. Maybe you have to make the personal effort to learn about the sky to truly get excited about astronomy.

If you are reading this book, then you are about to make that personal step. Inside the pages of *The Complete Idiot's Guide to Astronomy, Third Edition*, is all of the information you need to slake your thirst for astronomical knowledge. From the solar system to the most distant reaches of our universe, every kind of object is discussed, including what we know about it and how we know what we know, as well as the implications of this knowledge. Current results are presented in easily understood ways with special additional information set off from the rest of the material. Striking images and pictures from telescopes in space and on the ground—including color images for the first time in this edition—show you what you cannot see with your eye, the detailed beauty of the heavens.

If a book like *The Complete Idiot's Guide to Astronomy, Third Edition*, had been around when I first began to study astronomy, I probably would have done better in my college classes. This is no joke! A lot of the information in this book is cutting-edge stuff. Even some researchers might not know some of the information in these pages. Impress your astronomer friends (or your regular friends, professional astronomers are pretty rare) at parties, or, if you are a younger reader, your science teacher, by reciting some of the new results contained in this book.

Here is just one example. Chapter 21 includes information about the results obtained from a space telescope that flew in 2003 called the *Wilkinson Microwave Anisotropy Probe* (just the name of the satellite will impress most people). This satellite confirmed results from earlier observations that we live in a universe that is geometrically flat and has a special density (you'll have to learn the details by reading the chapter!). These observations combined with new observations (also discussed in Chapter 21) of distant supernovas tell us that all the stars and galaxies we see make up only a small fraction of the total mass of the universe and that our universe is expanding faster and faster over time. We are but dust in an ever-quickening wind according to these new results. Including new and cutting-edge results in a book for novice astronomers is a great thing and a unique value in this volume.

I work in Washington, D.C., trying to advocate for increased spending for basic research, especially astronomy. I am an astronomical lobbyist. Scientific and technical issues often intimidate members of Congress. "Get me an expert," they often say, "I wasn't trained as a scientist." But they are quite happy to speak at length on social issues, taxation, the economy, international relations, and so on, even if they weren't trained in those fields (most of them are lawyers).

This is a common feeling in our country. Science is somehow especially difficult or only understood through considerable effort by very smart people. Nothing could be further from the truth. Non-scientists can easily understand cutting-edge results and everyone should know the basics. This book makes the hard stuff easy to understand and the easy stuff easier to understand. You'll see.

I've known one of the authors, Chris De Pree, since he was wearing professional "diapers." We shared an office while working on our doctorate degrees and made home-brewed beer on the weekends (most of the time it tasted good). Aside from his somewhat messy habits, poor taste in music, ability to whistle perfectly out of tune, and small grunting noises he makes when concentrating, he was a good office mate. He is a phenomenal author and I have to say that being able to write this foreword has been a great honor. Plus, I got a pre-publication copy of the book for free! Chris's editors and co-author have made sure none of his messy habits remain in this volume and that all of his creativity and expertise are front-and-center. I am sure you will enjoy reading its pages as much as I have.

—Kevin B. Marvel, Ph.D.

Deputy executive officer, American Astronomical Society

Introduction

You are not alone.

Relax. That statement has nothing to do with the existence of extraterrestrial life—though we will get around to that, too, way out in Chapter 22. For the present, it applies only to our mutual interest in astronomy. For we (the authors) and you (the reader) have come together because we are the kind of people who look up at the sky a lot and have all kinds of questions about it. This habit hardly brands us as unique. Astronomy, the scientific study of matter in the universe, is among the most ancient of human studies. The very earliest scientific records we have—from Babylon, from Egypt, from China—all concern astronomy.

Recorded history spans about 5,500 years. The recorded history of astronomy starts at the beginning of that period. Humans have been sky watchers for a very, very long time.

And yet astronomy is also among the most modern of sciences. Although we possess the collected celestial observations of some 50 centuries, almost all that we know about the universe we have learned in the century just ended, and an enormous amount of essential knowledge has been gathered since the development of radio astronomy in the 1950s. In fact, the lifetime of any reader of this book, no matter how young, is filled with astronomical discoveries that merit being called milestones. Think it was a pretty big deal when Copernicus, in the early sixteenth century, proposed that the Sun, not Earth, was at the heart of the solar system? Well, did you know that a Greek astronomer actually proposed the same idea nearly 2,000 years earlier? His pitch just wasn't as good.

Astronomy is an ancient science on the cutting edge. Great discoveries were made centuries ago; great discoveries are being made today. And great leaps forward in astronomical knowledge have often followed leaps forward in technology: the invention of the telescope, the invention of the computer, and the development of fast, cheap computers. So much is being learned every day that we've been asked to bring out a revised edition of this book, the third in only five years. And even more recent discoveries will be on the table by the time you read this latest edition.

Yet you don't have to be a government or university scientist with your eager fingers on millions of dollars' worth of equipment to make those discoveries. For if astronomy is both ancient and advanced, it is also universally accessible: up for grabs.

The sky belongs to anyone with eyes, a mind, imagination, a spark of curiosity, and the capacity for wonder. If you've also got a few dollars to spend, a good pair of binoculars or a telescope makes more of the sky available to you. (Even if you don't want to spend the money, chances are your local astronomy club will let you use members' equipment if you come and join them for a cold night under the stars.) And if you have a computer and Internet connection available, you—yes, you—have access to much of the information that those millions of dollars in government equipment produce: images from the world's great telescopes and from a wealth of satellite probes, including the *Hubble Space Telescope* and the *Mars Global Surveyor*. This information is all free for the downloading. (See Appendix E for some starting points in your online searches.)

We are not alone. No science is more inclusive than astronomy.

Nor is astronomy strictly a spectator sport. You don't have to peek through a hole in the fence and watch the game. You're welcome to step right up to the plate. Many new comets are discovered by astronomy buffs and backyard sky watchers as well as Ph.D. scientists in domed observatories. Most meteor observations are the work of amateurs. You can even get in on such seemingly esoteric fields as radio astronomy and the search for extraterrestrial intelligence.

We'd enjoy nothing more than to help you get started on your journey into astronomy. Here's a map.

How This Book Is Organized

Part 1, "Our Place in Space," orients you in the evening sky and presents a brief history of astronomy.

Part 2, "Worlds Without End," begins with a visit to our nearest neighbor, the Moon, and then ventures out into the rest of the solar system for a close look at the planets and their moons, as well as such objects as asteroids and comets.

Part 3, "To the Stars," begins with our own Sun, taking it apart, showing how it works, and providing instructions for safely viewing it both day to day and during an eclipse. From our Sun, we venture beyond the solar system to the other stars and learn how they are meaningfully observed. In the end, we explore the very strange realm of neutron stars and black holes.

Part 4, "Way Out of This World," pulls back from individual stars to take in entire galaxies, beginning with our own Milky Way. We learn how astronomers observe, measure, classify, and study galaxies and how those galaxies are all rushing away from us at incredible speed. The section ends with the so-called active galaxies, which emit unimaginably huge quantities of energy and can tell us much about the origin and fate of the universe.

Part 5, "The Big Questions," asks how the universe was born (and offers the Big Bang theory by way of an answer), then if (and how) the universe will end. Finally, we explore the possibilities of extraterrestrial life and even extraterrestrial civilizations.

At the back of the book, you'll find a series of appendixes that define key terms, list upcoming eclipses, catalog the constellations, provide the classic Messier Catalog of deep-space objects that amateurs can readily observe, and list sources of additional information, including great astronomy websites.

Extras

In addition to the main text and illustrations of *The Complete Idiot's Guide to Astronomy, Third Edition,* you'll also find other types of useful information, including definitions of key terms, important statistics and scientific principles, amazing facts, and special subjects of interest to sky watchers. Look for these features:

Close Encounter

In these boxes, you'll find discussions elaborating on important events, projects, issues, or persons in astronomy.

Astronomer's Notebook

This feature highlights important statistics, scientific laws and principles, measurements, and mathematical formulas.

Star Words

These boxes define some key terms used in astronomy.

Astro Byte

Here are some startling astronomical facts and amazing trivia. Strange—but true!

Acknowledgments

I would like to acknowledge two scientists who had a big influence on my decision to become an astronomer, and one friend who has made being an astronomer much more fun. Dr. Jon Kolena taught me as an undergraduate physics major at Duke University when I was a senior; his course was the only undergraduate class in astronomy offered at the time. It was a challenging course with lots of problem solving, and it helped me to begin to understand the ways in which my physics training could be used to solve problems in astronomy. I had never considered a career in astronomy before taking his class, and it has been a great pleasure to see him at professional meetings since that time. Dr. Wayne Christiansen is a professor of astronomy at the University of North Carolina at Chapel Hill. He sent me as a green first-year graduate student to a summer institute at the National Radio Astronomy Observatory (NRAO) in Socorro,

New Mexico, in 1991. That brief trip convinced me that radio astronomy was an amazing field, and led to my becoming not only an astronomer, but a radio astronomer, and I thank him for that. And my office mate as a graduate student in Socorro was my friend and colleague Kevin Marvel, now with the American Astronomical Society. His humor, enthusiasm, friendship, and intelligence have been an inspiration, though we now live too far away from each other to brew beer together.

Finally, I would like to acknowledge the employees of the National Radio Astronomy Observatory. From my days in Socorro as a graduate student to my return trips as a professor to Socorro, New Mexico; Green Bank, West Virgina; and Charlottesville, Virginia; this dedicated group of astronomers, engineers, staff, and support personnel have been like an extended family.

Trademarks

All terms mentioned in this book that are known to be or are suspected of being trademarks or service marks have been appropriately capitalized. Alpha Books and Penguin Group (USA) Inc. cannot attest to the accuracy of this information. Use of a term in this book should not be regarded as affecting the validity of any trademark or service mark.

Part 1

Our Place in Space

We know this isn't your first night out and that you've certainly looked up at the sky before. Maybe you can find the Big Dipper and even Orion—or at least his Belt—but, for the most part, all the stars look pretty much the same to you, and you can't tell a star from a planet.

The first chapter of this part gets you started with the constellations. The second chapter says a bit about the history of astronomy. The third introduces the telescope, and the fourth and fifth take you through the electromagnetic spectrum and beyond visible light.

Naked Sky, Naked Eye: Finding Your Way in the Dark

In This Chapter

◆ What you can see with your naked eye

◆ The celestial sphere

◆ Orienting yourself among the stars

◆ Celestial coordinates and altazimuth coordinates

◆ Identifying constellations: how and why

Want to make a movie on an extraterrestrial theme? Hollywood has been using space as a backdrop for quite some time, and it's still big box office these days. Audiences continue to be thrilled by special effects: blazing comets, flaming meteors, brightly banded planets, and strange, dark moons. Just be prepared to spend upward of $100 million to make your film. Those special effects don't come cheap, and today's moviegoers are spoiled by one dazzling spectacle after another. Whatever did people do for excitement before *2001: A Space Odyssey, Independence Day, The Fifth Element, Contact,* and the latest sequels to *Star Wars* and *Star Trek?*

They looked at the sky.

This chapter will tell you what they saw.

Sun Days

We've become jaded—and a bit spoiled—by the increasingly elaborate and costly special effects in today's sci-fi flicks, but none of us these days is nearly as spoiled as the sky most of us look at.

Imagine yourself as one of your ancestors, say 10,000 years ago. Your reality consists of a few tools, household utensils, perhaps buildings (the city-states were beginning to appear along the Tigris River), and, of course, all that nature has to offer: trees, hills, plants, rivers, streams—and the sky.

The sky is the biggest, greatest, most spectacular object you know. During the day, the sky is crossed by a brightly glowing disk from which all light and warmth emanate. Announced in the predawn hours by a pink glow on the eastern horizon, the great disk rises, then arcs across the sky, deepening toward twilight into a ruddy hue before slipping below the horizon to the west. Without electric power, your working hours are largely dictated by the presence of the Sun's light.

Flat Earth, Big Bowl

As the Sun's glow fades and your eyes become accustomed to the night, the sky gradually fills with stars. Thousands of them shimmer blue, silvery white, some gold, some reddish, seemingly set into a great dark bowl, the *celestial sphere,* overarching the flat Earth on which you stand.

Thousands of stars in the night sky, you say?

Maybe that number has brought you back through a starlit 10,000 years and into the incandescent lamp light of your living room or kitchen or bedroom or wherever you are reading this: *"I've* never seen thousands of stars!" you protest.

> **Star Words**
>
> The **celestial sphere** is an imaginary sphere that we picture surrounding Earth upon which the stars are fixed. Some ancient cultures believed such a sphere (or bowl) actually existed. Today, however, astronomers use the concept as a way to map the location of stars relative to observers on Earth.

We said earlier that, from many locations, our sky is spoiled. The sad fact is that, these days, fewer and fewer of us can see anything like the 6,000 or so stars that *should* be visible to the naked eye on a clear evening. Ten thousand years ago, the night sky wasn't lit up with the *light pollution* of so many sources of artificial illumination. Unless you sail far out to sea or travel to the high, dry desert of the American Southwest, you might go through your entire life without *really* seeing the night sky, at least not the way our ancestors saw it.

> **Star Words**
>
> **Light pollution** is the result of photons of light that go up instead of down. Light that goes *down* (from fixtures) illuminates the ground. Light that goes *up* makes the stars harder to see. You can contact your local astronomy club to find ways to combat light

Man in the Moon

Even in our smog- and light-polluted skies, however, the Moon shines bright and clear. Unlike the Sun, which appears uniform, the surface of the Moon has details we can see, even without a telescope. Even now, some three decades after human beings walked, skipped, and drove on the Moon and even hit a golf ball across the lunar surface, the Moon holds wonder. Bathed in its silver glow, we might feel a connection with our ancestors of 10 millennia ago. Like them, we see in the lunar blotches the face of the "Man in the Moon."

> **Close Encounter**
>
> For untold generations, people have discerned a human face in the crater-scarred markings of the Moon. The Man in the Moon is sometimes interpreted as an old woman cooking. Among Native Americans, the face or faces in the Moon have sometimes been described as a frog charged with protecting the Moon from a bear who would otherwise swallow it. An ancient Scandinavian folktale speaks of Hjuki and Bill, perhaps the original Jack and Jill, who, carrying a pail of water, tumble down a hill as they run from their cruel father. They are rescued by the embrace of the Moon. For Scandinavian kids, the "Man in the Moon" is the image of Hjuki and Bill, complete with pail.

If the face of the Moon presented a puzzle to our ancestors, they were also fascinated by the way the Moon apparently changed shape. One night, the Moon might be invisible (a "new moon"); then, night by night, it would appear to grow (wax), becoming a crescent; and, by one week later, be a quarter moon (which is a *half* moon in shape).

new moon → wax ¹ʷᵏ⟶ quarter moon ⟶ gibbous
(grow)
crescent (half)

Star Words

Gibbous is a word from Middle English that means "bulging"—an apt description of the Moon's shape between its quarter phase and full phase.

Through the following week, the Moon would continue to wax, entering its *gibbous* phase, in which more than half of the lunar disk was seen. Finally, two weeks after the new moon, all of the lunar disk would be visible: The full moon would rise majestically at sunset. Then, through the next two weeks, the Moon would appear to shrink (wane) night after night, passing back through the waning gibbous, third quarter, and waning crescent phases, until it became again the all-but-invisible new moon.

Neil Armstrong took this picture of fellow astronaut "Buzz" Aldrin, about to join him on the surface of the Moon, July 20, 1969.

(Image from arttoday.com)

The cycle takes a little more than 29 days—a month, give or take a day—and it should be no surprise that the word "month" is derived from the word "moon." In fact, just as our ancestors learned to tell the time of day from the position of the Sun, so they measured what we call weeks and months by the lunar phases. The lunar calendar is of particular importance in many world religions, including Judaism and Islam. For those who came before us, the sky was more than something to marvel at. It could also be used to guide and coordinate human activity. The ancients became remarkably adept at using the heavens as a great clock and calendar. Nature was not kind, though, in giving us units of time. The day, month, and year are not evenly divisible into one another, and there are 365.25 days in a year, a fact that has caused much consternation to calendar makers over the centuries.

The phases of the Moon. The globe in the center is Earth. The inner circle shows how the sunlight illuminates the Moon as it orbits Earth. The outer circle shows how the Moon appears from Earth.

(Image from the authors' collection)

Lights and Wanderers

Ten thousand years ago, family time at night wasn't occupied with primetime sitcoms followed by the news and *The Daily Show.* Our ancestors weren't glued to television screens, but presumably to the free show above, the celestial sphere. Early cultures noticed that the bowl above them rotated from east to west. They concluded that what they were seeing was the celestial sphere—which contained the stars—rotating, and not the individual stars. All the stars, they noticed, moved together, their positions relative to one another remaining unchanged. (That the stars "move" because of *Earth's* rotation was an idea that would be millennia in the making.)

The coordinated movement of the stars was in dramatic contrast to something else the ancient sky watchers noticed. Although the vast majority of stars were clearly fixed in the rotating celestial sphere, a few—the ancients counted five—seemed to meander independently, yet regularly, across the celestial sphere. The Greeks called these five objects *planetes,* "wanderers," and, like nonconformists in an otherwise orderly society, the wanderers would eventually cause trouble. The details of their movement would bring the entire heavenly status quo into question and, ultimately, the whole celestial sphere would come crashing down.

Celestial Coordinates

But we're getting ahead of our story. Later, you'll find out why we no longer believe that the celestial sphere represents reality; however, the notion of such a fixed structure holding the stars is still a useful model for us moderns. It helps us to communicate with others about the positions of the objects in the sky. We can orient our gaze into the heavens by thinking of the point of sky directly above Earth's North Pole as the north celestial pole, and the point below the South Pole as the south celestial pole. Just as Earth's equator lies midway between the North and South Poles, so the celestial equator lies equidistant between the north and south celestial poles. Think of it this way: If you were standing at the North Pole, then the north celestial pole would be directly overhead. If you were standing at the equator, the north and south celestial poles would be on opposite horizons. And if you were standing at the South Pole, the south celestial pole would be directly overhead.

Astronomers have extended to the celestial sphere the same system of latitude and longitude that describes earthly coordinates. The lines of latitude, as you might recall from geography, run parallel with the equator and measure angular distance north or south of the equator. On the celestial sphere, *declination* corresponds to latitude and measures the angular distance above or below the celestial equator. Although earth-bound latitude is expressed in degrees north or south of the equator (Philadelphia, for instance, is 40 degrees north), celestial declination is expressed in degrees + (above) or – (below) the celestial equator. The star named Betelgeuse, for example, is at a declination of +7 degrees, 24 minutes (1 minute = $\frac{1}{60}$ degree).

Declination is analogous to Earthly latitude. The declination of a star seen directly above Earth's equator would also be at the celestial equator—that is, 0 degrees. A star at the north celestial pole (that is, directly over Earth's North Pole) would be +90 degrees. At the south celestial pole, it would be –90 degrees. In the latitudes of the United States, stars directly overhead have declinations in the +30 to +40 degree ranges. The Bradley Observatory at Agnes Scott College is at a latitude of 33 deg, 45 min, 55.84 sec. That means that in Decatur, GA (just outside of Atlanta), the North Star (Polaris) is about 34 degrees above the northern horizon.

> **Star Words**
>
> **Declination** is the angular distance (distance expressed as an angle rather than in absolute units, such as feet or miles) north or south of the celestial equator. It is akin to lines of latitude on Earth.

> **Star Words**
>
> **Right ascension** is a coordinate for measuring the east-west position of objects in the sky.

On a globe, the lines of longitude run vertically from pole to pole. They mark angular distance measured east and west of the so-called prime meridian (that is, 0 degrees), which by convention and history has been fixed at Greenwich Observatory, in Greenwich, England. On the celestial sphere, *right ascension* (R.A.) corresponds to longitude. Although declination is measured as an angle (degrees, minutes, and seconds), right ascension is measured in hours, minutes, and seconds, increasing from west to east, starting at 0. This zero point is taken to be the position of the Sun in the sky at the moment of the vernal equinox (we'll get to this a bit later). Because Earth rotates once approximately every 24 hours, the same objects will return to their positions in the sky approximately 24 hours later. After 24 hours, Earth has rotated through 360 degrees, so that each hour of R.A. corresponds to 15 degrees on the sky (because 360 divided by 24 equals 15).

Bradley Observatory is the college observatory of Agnes Scott College in Decatur, Georgia. Like many observatories across the country, it is home to a telescope dome and a planetarium.

(Image from the authors' collection)

If the celestial poles and the celestial equator are projections of Earthly coordinates (the poles and the equator), why not simply imagine *right ascension* as projections of lines of longitude?

There are good reasons why we don't. Think of it this way: The stars in the sky above your head in winter are different than those in summer. That is, in the winter we see the constellation Orion, for example, but in the summer, Orion is gone, hidden in the glare of a much closer star, the Sun. Well, although the stars above you are changing daily, your longitude (in Chicago, for example) is not changing. So the coordinates of the stars cannot be fixed to the coordinates on the surface of Earth. As we'll see later, this difference comes from the fact that in addition to spinning on its axis, Earth is also orbiting the Sun.

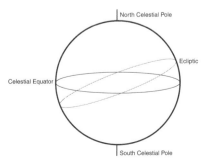

The celestial sphere is shown with the following features labeled: the celestial equator, the ecliptic (the apparent path of the Sun), and the north and south celestial poles.

Measuring the Sky

The true value of the celestial coordinate system is that it gives the absolute coordinates of an object so that two observers, anywhere on Earth, can direct their gaze to the exact same star. When you want to meet a friend in the big city, you don't tell her that you'll get together "somewhere downtown." You give precise coordinates: "Let's meet at the corner of State and Madison streets." Similarly, the right ascension and declination that astronomers use tell them (and you) precisely where in the sky to look.

The celestial coordinate system can be confusing for the beginning sky watcher. However, an understanding of this system of coordinates can help the novice observer locate the North Star, know approximately where in the sky to look for planets, and know where in the sky to look for Messier objects, for example, the coordinates of which are given in right ascension and declination.

There is a simpler way to measure the location of an object in the sky as observed from your location at a particular time. It involves two angles, azimuth and altitude. You can use angles to divide up the horizon by thinking of yourself as standing at the center of a circle that's flat on the ground around you. A circle can be divided into 360 degrees (and a degree can be subdivided into 60 minutes, and a minute sliced into 60 seconds). When you decide which direction is 0 degrees (the convention is to take due north as 0 degrees), you can measure, in degrees, precisely how far an object is from that point. Now that you have taken care of your horizontal direction, you can fix your vertical point of view by imagining an upright half circle extending from horizon to horizon. Divide this circle into 180 degrees, with the 90-degree point directly overhead. Astronomers call this overhead point the _zenith_.

Altitude and azimuth are the coordinates that, together, make up the altazimuth coordinate system, and, for most people, they are quite a bit easier to use than celestial coordinates of declination and right ascension. An object's _altitude_ is its angular distance above the horizon, and its compass direction, called _azimuth_, is measured in degrees increasing clockwise from due north. Thus east is at 90 degrees, south at 180 degrees, and west at 270 degrees.

Star Words

Altazimuth coordinates are **altitude** (angular distance above the horizon) and **azimuth** (compass direction expressed in angular measure east of due north).

Altazimuth coordinates, while perhaps more intuitive than the celestial coordinate system, do have a serious shortcoming. They are valid only for your location on Earth at a particular time of day or night. In contrast, the celestial coordinate system is universal because it moves with the stars in the sky. For this reason, star catalogs list the right ascension and declination of objects, not their altitude and azimuth coordinates, which are always changing!

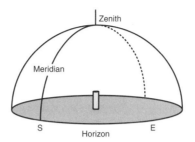

An observer in the altazimuth system, showing the local horizon, meridian, and zenith. The problem with this system is that every observer has a different local horizon.

"I Am Crushing Your Head!"

In a television show (still being rerun) called *Kids in the Hall*, one character would look at people far away through one eye and pretend to crush their heads between his thumb and forefinger. If you try this trick yourself, you'll notice that people have to be at least five or so feet away for their heads to be small enough to crush. Their heads don't actually get smaller, of course, just the angular size of the heads does. In fact, you can use this same trick to crush cars (if sufficiently distant) or planes flying overhead. It's all because of the fact that as things get more distant, they appear smaller—that is, their *angular size* is reduced.

The surface of Earth is real and solid. You can easily use absolute units such as feet and miles to measure the distance between objects. The celestial sphere, however, is an imaginary construct, and we don't know the distances between us and the objects we see. Fortunately, to locate objects in the sky, we don't need to know their distances from us. We get that information in other ways, which we will discuss later in this book. Now, from our perspective on Earth, two stars might appear to be separated by the width of a finger held at arm's length when they are actually many trillions of miles distant from each other. You could try to fix the measurement between two stars with a ruler, but where would you hold the measuring stick? Put the ruler close to your eye, and two stars might be a quarter-inch apart. Put it at arm's length, and the "distance" between those same two stars might grow to several inches. Obviously, we need a different approach.

Astronomers use concepts called *angular size* and *angular separation* to discuss the apparent size in the sky or apparent distance between two objects in the sky. For example, if two objects were on opposite horizons, they would be 180 degrees apart. If one were on the horizon and the other directly overhead, they would be 90 degrees apart. You get the picture. Well, a degree is made up of even smaller increments. One degree is made up of 60 minutes (or *arcminutes*), and a minute is divided into 60 seconds (*arcseconds*).

Let's establish a quick and dirty scale. The full moon has an angular size of half a degree, or 30 arcminutes, or 1,800 arcseconds (these are all equivalent). Now that you know the full moon is about half a degree across, you can use its diameter to gauge other angular sizes.

Star Words

Angular size and angular separation are size and distance expressed as angles on the sky rather than as absolute units (such as feet or miles). Because many of these measurements are less than a full degree, you should know a degree is made up of 60 **arcminutes** and an arcminute of 60 **arcseconds**.

You can use your hand to estimate angles greater than a half-degree. Look at the sky. Now extend your arm in front of you with your wrist bent so that the back of your hand is facing you. Spread your thumb, index finger, and pinky fully, and fold your middle finger and ring finger down so you can't see them. Voilà—a *handy* measuring device! The distance from the tip of your thumb to the tip of your index finger is about 20 degrees (depending on the length of your fingers!). From the tip of your index finger to the tip of your pinky is 15 degrees; and the gap between the base of your index finger and the base of your pinky is about 10 degrees.

Astronomer's Notebook

The "smallest" celestial object the human eye can resolve is about 1 arcminute across. The largest lunar craters are about 2 arcminutes across, and without new technologies such as adaptive optics, separating objects that are 1–2 arcseconds apart is impossible (at least at optical wavelengths) from all but the best sites on Earth. This difficulty is due to atmospheric turbulence and is a limitation of current ground-based optical observing.

Some of the world's best ground-based optical telescopes are employing a new technology called "adaptive optics," which involves distorting the surfaces of small mirrors in the optical path of a telescope in real time to correct for fluctuations in Earth's atmosphere. With adaptive optics, even ground-based telescopes can obtain resolution better than 1 arcsecond, an accomplishment that used to be possible only for space-based optical telescopes such as the *Hubble*.

Celestial Portraits

Well, now that you're standing there with your arm outstretched and your head full of angles, what can you do with this wealth of information?

You now have some rough tools for measuring separations and sizes in the sky, but you still need a way to *anchor* your altazimuth measurements, which, remember, are relative to where the person doing the measuring happens to be standing on Earth. We need the celestial equivalent of landmarks.

Fortunately for us, our ancestors had vivid imaginations.

Human brains are natural pattern makers. We have all seen elephants and lions masquerading as clouds in the sky. Present the mind with the spectacle of 3,000 randomly placed points of light against a sable sky, and, before you know it, it will start "seeing" some pretty incredible pictures. The *constellations*—arbitrary formations of stars that are perceived as figures or designs—are such pictures, many of them named after mythological heroes, whose images (in the western world) the Greeks created by connecting the dots.

By the second century C.E., Ptolemy (whom you'll meet a bit later) listed 48 constellations in his *Almagest*, a compendium of astronomical knowledge. Centuries later, during the late Renaissance, more constellations were added, and a total of 88 are internationally recognized today. We cannot say that the constellations were really discovered, because they don't exist except in the minds of those who see them. Grouping stars into constellations is an arbitrary act of the imagination and to present-day astronomers is merely a convenience. In much the same way that states are divided into counties, the night sky is divided into constellations. The stars thus grouped generally have no physical relationship to one another. Nor do they necessarily lie at the same distance from Earth; some are much farther from us than others. But, remember, we simply imagine, for the sake of convenience, that they are embedded in the celestial sphere.

Astronomer's Notebook

Of the 88 constellations, 28 are in the northern sky and 48 are in the southern sky. The remaining dozen lie along the ecliptic—the circle that describes the path that the Sun takes in the course of a year against the background stars. This apparent motion is actually due to Earth moving around the Sun. These 12 constellations are the zodiac, familiar to many as the basis of the pseudoscience (a body of lore masquerading as fact verified by observation) of astrology. All but the southernmost 18 of the 88 constellations are at least sometimes visible from part of the United States.

If the constellations are outmoded figments of the imagination, why bother with them?

The answer is that they are convenient (not to mention poetic) celestial landmarks. We all use landmarks to navigate on land. "Take a right at Hank's gas station," you might tell a friend. What's so special about that particular gas station? Nothing—until you endow it with significance as a landmark. Nor was there anything special about a group of physically unrelated stars—until they were endowed with significance. Now these constellations can help us find our way in the sky and to the casual night observer are more useful than either the celestial or altazimuth coordinate system.

Astro Byte

In an age when so many astronomical objects, of necessity, are referred to by rather cold catalog names (NGC 4258, W49A, K3-50A, to name a few), it is pleasing that we can still refer to some objects by their brightness within a given constellation. Cygnus X-1 is a famous x-ray source and black-hole candidate in the constellation of Cygnus, the swan.

The Dippers First

Almost everybody knows the Big Dipper and maybe the Little Dipper, too. Actually, neither Dipper is a constellation, but they are subsets of two other constellations, Ursa Major and Ursa Minor, the big and little bears (official constellation names are in Latin). Such generally recognizable subgroups within constellations are called *asterisms*. The Big Dipper is not only bright, but it is easy to find in the northern sky in all seasons except fall, when it is close to the horizon. It might interest you to know that you'll find the Big Dipper between 11 and 14 hours right ascension and +50 to +60 degrees declination. Using your hand to estimate the Big Dipper's angular size, you'll see that it's about 25 degrees across the bowl to the end of the handle. But the really important thing is that its seven bright stars form a pattern that really does look like a dipper.

And that's what's so handy about asterisms. They are often simpler, brighter, and more immediately recognizable than the larger, more complex constellations of which they are a subset. They will help you to find the constellations with which they are associated and generally help to orient you in the sky.

Seafarers and other wanderers have long used the Big Dipper as a navigational aid. If you trace an imaginary line between the two stars that mark the outer edge of the Big Dipper's bowl and extend that line beyond the top of the bowl, it points to Polaris, the North Star, about 25 degrees away. Polaris is very nearly at the north celestial pole (about 1 degree off), which means that it appears to move very little during the course of the night, and travelers have always used it as a compass. During the decades before the American Civil War, many Southern slaves escaped to the free North by following the North Star. For sky explorers, the combination of the Big Dipper and Polaris provide a major landmark useful in locating other constellations. One of those constellations—actually, it's another asterism— is the Little Dipper. Dimmer and smaller than the Big Dipper, it would be harder to find, except that Polaris, which we've just located, is at the tip of its handle. Like its big brother, this asterism consists of seven stars.

Star Words

An **asterism** is an arbitrary grouping of stars within or associated with a constellation, which are perceived to have a recognizable shape (such as a Teapot or Orion's Belt) and, therefore, readily serve as celestial landmarks.

The Stars of Spring

Let's look at a few of the highlights of each season's sky.

With the arc of our Galaxy (the Milky Way) low and heading toward the western horizon, the spring sky offers fewer bright stars than any other season. This isn't necessarily a bad thing, because it makes identifying the three bright ones that much easier. Some 45 degrees south of the Big Dipper's bowl is the constellation Leo. If you can't quite pick out Leo, you might find it easier to identify the asterism called the Sickle, a kind of backwards question mark that forms Leo the Lion's mane. At the base of the Sickle is the bright star Regulus.

Arcturus, another bright star of spring, can be located by extending the curve of the Big Dipper's handle 35 degrees southward.

Yellow-orange in color, Arcturus is the brightest star of the constellation Boötes, the charioteer.

Now extend the Big Dipper handle's curve beyond Arcturus, and you will find Spica (in Virgo), the third bright star of spring. In vivid contrast to the warm hue of Arcturus, Spica is electric blue. You'll find out in Chapter 13 that the color of a star actually tells you about its surface temperature. It can be quite a thrill to look at different stars in the sky and be able to "take their temperatures" simply from their colors.

> **Astro Byte**
>
> "Arc to Arcturus" is a handy mnemonic often taught to astronomy students to help them easily locate Arcturus. Following the arc of the Big Dipper's handle leads to this bright star.

Summer Nights

Summer offers four bright stars. Three of them form a distinct right triangle called the summer triangle. Vega is at the triangle's biggest angle. Vega is also the brightest star in the constellation Lyra, the Lyre (or harp). South and east of Vega is the second brightest star in the triangle, Altair, which is in the constellation Aquila, the Eagle. Deneb is the third star of the summer triangle, and is located in Cygnus, the Swan. Deneb is also part of the prominent asterism, the Northern Cross, and is the brightest star in that group.

Take a good long look at Deneb. Bright as it is—fourth brightest in the summer sky—it is one of the most distant stars visible to the naked eye (over 3,000 light-years away), over 100 times more distant than Vega and many hundreds of times farther than Alpha Centauri, our closest stellar companion (other than the star we call the Sun).

If you are sufficiently far from sources of light pollution, and the night is clear and dry, you might notice that the Northern Cross lies within a kind of hazy band stretching across the sky. This band is the Milky Way, our own Galaxy, which you will explore in Chapter 16 and whose haze is the light of more than 100 billion stars.

Two other major summer constellations should not be missed. Sagittarius, the archer, and Scorpius, the scorpion, are found low in the southern sky about 30 degrees below the celestial equator. You can locate Scorpius by finding the fourth bright star of the summer sky, Antares, unmistakable for the red hue that gives it its name, which means "rival of Mars." If Scorpius is not below the horizon (and therefore out of sight), you should recognize its fishhook-shaped scorpion's tail.

One hour R.A. (15 degrees) east of Scorpius is Sagittarius. You might better recognize it by two asterisms within it: the Teapot, which looks as if it pours out on the tail of nearby Scorpius, and the Milk Dipper, called this because its dipper shape seems to dip into the milkiest part (thickest star cloud) of the Milky Way. As you will see, there is a reason that the Milky Way is thicker here, bulging slightly. Sagittarius is the direction of the center of our own Galaxy.

Fall Constellations

In the fall, the constellation Pegasus, winged horse of Greek mythology, is easy to locate. If you find it hard to imagine connecting the stars to trace out the horse, look for the highly recognizable asterism associated with Pegasus called the Great Square. At southern latitudes, by about 10 P.M. in early October, it should be directly above you. The four stars marking out its four corners aren't terribly bright, but the other stars in that area of the sky are fairly dim, so the figure should stand out clearly. The eastern side of the Great Square also coincides with the 0 marking from which the hours of right ascension start, increasing to the east.

Astro Byte

There is an abundance of stars that still retain their Arabic names, a testament to the many contributions of Arab astronomers: Aldebaran, Mizar, Alcor, and Betelgeuse are a few of these stars.

Some 20 degrees west and 5 degrees south of Markab, the star that marks the Great Square's southwest corner, is Enif, the brightest star in Pegasus. Its name means "the horse's mouth," and between Markab and Enif is the horse's neck. Look to the Great Square's northeast corner for the star Alpheratz, which is not part of Pegasus, but part of Andromeda, the maiden in chains.

If you trace a line from Alpheratz through Markab, continuing about 40 degrees southwest of Markab, you'll find the zodiacal constellation Capricornus, Capricorn, or the sea goat. Capricorn is distinguished by its brightest star, the brilliant Deneb Algiedi.

Return to the Great Square. About 20 degrees east of it, you'll find another zodiacal constellation, Aries, the ram. This grouping is easy to identify, because it is marked by two fairly bright stars a mere 5 degrees apart.

Last to rise in the sky of fall is Perseus, slayer of snake-haired Medusa and other monsters of Greek mythology. About 45 degrees up in the northeast, it lies across the Milky Way and is marked by its brightest star, Mirfak.

Winter Skies

Winter nights, with the bright arc of the Milky Way overhead, offer more bright stars than are visible at any other season: Sirius, Capella, Rigel, Procyon, Aldebaran, Betelgeuse, Pollux, and Castor. Brightest and most readily recognizable of the winter constellations is Orion, the hunter, which spans the celestial equator and sports the heavens' second most familiar asterism (after the Big Dipper): Orion's Belt, three closely spaced bright stars in a line 3 degrees long. The star Rigel, brightest in the Orion constellation, marks the hunter's foot, 10 degrees below and to the west of Orion's Belt. About the same distance and direction above the Belt is Betelgeuse, a reddish star, whose name is Arabic for "armpit of the giant." And that is precisely what Betelgeuse marks: Orion's armpit. Another star, Bellatrix, marks the shoulder of Orion's arm holding his shield, which is an arc of closely spaced, albeit dim, stars. Suspended from Orion's Belt is a short sword, the middle "star" of which is actually a region where stars are being born. (We will discuss the Orion nebula and other regions like it in Chapter 14.)

Saiph is Orion's eastern leg. About 15 degrees to the southeast of this star is Sirius, called the Dog Star, because it is in the constellation Canis Major, the great dog. Sirius is the brightest star in the heavens.

To the northeast of Orion you will readily see a pair of bright stars close together. These are Castor and Pollux, the twins, which represent the two heads of the constellation Gemini. Moving in an arc to the northwest of Castor and Pollux, you should see another bright star, this one with a distinctly yellow-gold color. It is called Capella, which means "little she-goat," and the ancients thought the star was the color of a goat's eye. Capella is in the constellation Auriga, the Goatherd.

Return to Orion. Just to the northwest of his shield, you will find Taurus, the bull, which is marked by Aldebaran, a bright orange star that forms the constellation's bull's eye. And in the shoulder of the bull is the asterism known as the Pleiades, one of the few apparent groupings of stars that are actually close to one another, massive stars having recently formed. Early sky watchers imagined Taurus eternally charging the shield of Orion, who stood eternally poised to strike the animal with his upraised club.

It is admittedly difficult to imagine the bull in Taurus, although you might at least be able to discern a V-shaped asterism called the Hyades, which is the bull's mouth.

Do Constellations Matter?

Enjoy the constellations. The pleasures of getting to know them can occupy a lifetime, and it can be fun pointing them out to your friends and family. You will also find familiarity with them useful for quickly navigating the heavens. But you won't be hearing a lot more about the constellations in this book. Recognizing them as the products of human fantasy and not the design of the universe, modern astronomy has only limited use for them. But still we hear the echoes of mythology in modern discoveries. The first-known multiple-planet extrasolar planetary system—in other words, a "solar" system beyond our own solar system—orbits the star Upsilon Andromeda. And the first confirmed extrasolar planet discovered around a Sunlike star? 51 Pegasus.

The Least You Need to Know

- ◆ For the ancients, even without telescopes, the night sky was a source of great fascination, which we can share.

- ◆ To view the sky meaningfully, you need a system of orienting yourself and identifying certain key features. Celestial coordinates and altazimuth coordinates offer two such systems.

- ◆ Astronomers use angular size and angular distance to describe the apparent sizes and separations of objects in the sky.

- ◆ Constellations are imaginative groupings of stars perceived as images, many of them influenced by Greek mythology; however, these groupings are arbitrary, reflecting human imagination rather than any actual relationships between those stars.

- ◆ Constellations are useful as celestial landmarks to help orient your observations.

Ancient Evenings:
The First Watchers

In This Chapter

- A look at ancient astronomy
- The most important early astronomers, the Greeks
- The solar system according to Ptolemy …
- … and according to Copernicus
- Kepler reduces Tycho Brahe's mountain of observations to three simple laws governing the solar system
- Galileo makes use of the telescope
- Newton explains why the Moon is just a big apple

One of the great attractions of astronomy is that it is so new and yet so old. Astronomy asks many questions that push the envelope of human knowledge, yet it is the most ancient of sciences. The Babylonians, who lived in southeastern Mesopotamia between the Tigris and Euphrates rivers (present-day southern Iraq from Baghdad to the Persian Gulf), are the first people we know of who actively studied the stars and planets. As early as 3000 B.C.E., they seem to have identified constellations and, sometime later, developed a calendar tied to the recurrence of certain astronomical events (they didn't yet have NCAA basketball tournaments to let them know it was springtime).

Astronomy was only one of the Babylonian areas of knowledge basic to civilization. From ancient Babylonia came the first system of writing, cuneiform; the earliest known body of law, the Code of Hammurabi; the potter's wheel; the sailboat; the seed plow; and even the form of government known as the city-state. And whenever people sought to bring order and understanding to their world, astronomy was part of the effort.

Archaic Astronomy

The Chinese weren't far behind the ancient Babylonians in making systematic observations of the heavens. Records show that the Chinese had arrived at the concept of a 365-day year, based on what appeared to be the Sun's annual journey across the background stars, prior to 2500 B.C.E.

A distinctly Western conception of the monster the ancient Chinese believed menaced the Sun during an eclipse.

(From Century Magazine, 1885)

Time, Space, and Harmony in China

How our ancestors accounted for what they observed says a lot about their beliefs at the time. For instance, the Chinese thought that an *eclipse* took place when a great dragon attacked the Sun, trying to consume it. Because a dragon is a beast, the Chinese reasoned, it might be susceptible to fear. So, in the midst of an eclipse, people would gather to shout, strike gongs, and generally make as much noise as possible. This always seemed to work, because sooner or later the dragon stopped eating the Sun. After all, it always emerged from the eclipse.

Because it was important to assemble as many people to make as much noise as possible, it was necessary to get advance warning of an eclipse. Accordingly, generations of

Chinese astronomers observed solar eclipses and discovered something they called the Saros, a cycle in which the Sun, Moon, and Earth are aligned in a particular way every 18 years and 11.3 days—more or less. Armed with knowledge of the Saros, the Chinese were able to predict most eclipses. Moreover, the Chinese *made records* of their astronomical observations. Astronomy was a way to measure the heavens and the passage of time, so that it might be possible to live in harmony with nature, which is something especially important in Chinese philosophies. Also, in terms of practical politics, an exact knowledge of the heavens aided rulers in establishing and maintaining their absolute authority by giving them a degree of control over precisely how time was measured and, therefore, how the important activities of state were ordered. Rulers could also interpret (and some times "predict") various heavenly phenomena as portents and signs affirming their authority.

Star Words

An **eclipse** is an astronomical event in which one body passes in front of another, so that the light from the occulted (shadowed) body is blocked. For example, the Sun, Moon, and Earth can align so that the Moon blocks the light from the Sun. This alignment produces a solar eclipse.

The Venus Tablet of Babylon

Back in Babylon, some time between 1792 and 1750 B.C.E., during the reign of Hammurabi (the Babylonian king who gave the world its first code of laws), the so-called Venus Tablet was inscribed, devoted to describing the behavior of that planet. Babylonian astronomers believed that the movements and positions of planets with respect to the constellations could influence the fate of kings and nations. This interest in the positions of planets as a portent of the future was one early motivation for careful study of the heavens.

Close Encounter

Using binoculars or a telescope, you might be able to see Venus in its phases, from thin crescent to full. During much of the year, the planet is bright enough to see even in daylight. Venus is closer to the Sun than we are, and that fact keeps it close to the Sun in the sky (never more than 47 degrees, or about a quarter of the sky from horizon to horizon). The best times to observe Venus are at twilight, just before the sky becomes dark, or just before dawn. The planet will be full when it is on the far side of the Sun from us and crescent when it is on the same side of the Sun as we are. With a telescope on a dark night you might be able to observe the "ashen light" phenomenon: When the planet is at quarter phase or less, a faint glow makes the unilluminated face of Venus visible.

It is easy to imagine why, of all the planets, brilliant Venus captured the attention of the Babylonians. It is the brightest celestial object after the Sun and the Moon, and, like the Moon, has distinct phases, as seen here.

(Image from arttoday.com)

The Celestial Pyramids of Egypt

While the Chinese, the Babylonians, and other ancient peoples used astronomical observations to help them rule and regulate the living, the ancient Egyptians used astronomy to help the dead find their place in the afterlife.

To think of the Egyptians is to think of the pyramids, the great tombs of the pharaohs. Prayers carved into the walls of pyramid chambers make reference to the stars and to the pharaoh's ascent into the sky among them. Texts inscribed on the monuments at Saqqara, Egypt, tell of the pharaoh joining the *circumpolar stars*, which neither rise nor set, and therefore live eternally. These texts also tell of the pharaoh's journey to the constellation Orion—identified by the Egyptians with Osiris, the eternally resurrected god.

Star Words

The **circumpolar stars** are those stars near the north celestial pole, that is, close to Polaris, the North Star. Depending upon your location on Earth, these stars might be directly overhead (if you live at the North Pole) or at some angle to the horizon.

Some archaeologists have even speculated that the pyramids themselves are observatories. Certainly the Great Pyramid, by far the largest of the 80 or so known pyramids along the Nile's west bank, is celestially aligned. Internal shafts or ducts point to the star Thuban, and other shafts point to Orion's Belt at certain times of the year, as if to indicate the afterlife destiny of the pharaoh, toward the deathless North Star (which neither rises nor sets) on the one hand, and toward the constellation associated with the eternally reborn god Osiris on the other.

Stonehenge and the New World

The ancient peoples of the Far East and the Middle East had no monopoly on the stars. The endlessly fascinating Stonehenge, built in stages between about 2800 B.C.E. and 1550 B.C.E. on the Salisbury Plain in Wiltshire, England, seems almost certainly to have been designed as a kind of astronomical observatory or, as some scholars have argued, a computer of astronomical phenomena. Various features of Stonehenge are aligned on the positions of the Sun and Moon where each rise and set on particular days known as *solstices* (the days of longest and briefest daylight, which begin the summer and winter, respectively). Thus it is widely agreed that Stonehenge was at least in part used for the keeping of a calendar.

The New World, too, has its celestially oriented ancient structures. Around 900 C.E., at Cahokia, in present-day southern Illinois, a Native American people known to us as the Mound Builders erected more than a hundred earthen mounds, the layout of which seem to mirror a concept of a cosmic plan. Farther south, the Maya of Mexico built magnificent stepped pyramids, clearly oriented toward the sunset at the winter solstice, as if to mark the annual "death" of the Sun.

The Maya (and later, the Aztecs) used celestial observation to formulate a calendar as accurate as that which the Chinese had developed. A host of North American, Mesoamerican, and South American Indian monuments reveal careful orientation toward specific astronomical events. For example, the great *kiva* (a subterranean ceremonial chamber perhaps dating from C.E. 700 to 1050) at Chaco Canyon, New Mexico, is oriented to mark the sunrise on the day of the summer solstice. It was built by the Anasazi, ancestors of the Zuni, Hopi, and Navajo.

Star Words

The **summer solstice** occurs on or about June 21, is the longest day in the Northern Hemisphere, and marks the beginning of summer. The **winter solstice** occurs on or about December 21 and, again in the Northern Hemisphere, is the shortest day and the start of winter. Many of our winter festivals (Chanukah and Christmas) have their roots in ancient celebrations of the winter solstice.

Astro Byte

The autumnal equinox occurs about September 22, and the vernal equinox about March 21, marking the beginning of fall and spring, respectively. On these dates, day and night are of equal duration. The Sun's apparent course in the sky (the ecliptic) intersects the celestial equator at these points.

What the Greeks Gave Us

More than any other ancient people, the Greeks left thorough written records of what they observed and what they thought, not only about the world around them, but the heavens above.

Earth Floats, Stars Burn

Anaximander (610–546/545 B.C.E.) is often called the founder of astronomy. He theorized that the world and everything in it were derived from an imperceptible substance he called the *apeiron* (unlimited), which was separated into various contrasting qualities and eventually differentiated into all matter, including Earth. Anaximander rejected what was then the prevailing notion that Earth was suspended from or supported by something in the heavens, holding instead that Earth floated freely in space at the center of the universe. As for the stars, they were fiery jets, and the Sun a chariot wheel whose rim was hollow and filled with fire.

Another Greek, Anaximenes of Miletus (active around 545 B.C.E.), theorized that what he called *aer* (air or vapor) was the most basic form of matter and was also the substance that formed the life spirit of animals, the soul of humankind, and the divine essence of the gods. When rarified, *aer* turned to fire, and the Sun, Moon, and stars were collections of rarified *aer*, masses of fire, set into a great crystal hemisphere that was the sky.

Anaximander and Anaximenes might no longer be household names, but a lot of us remember Pythagoras (ca. 580–ca. 500 B.C.E.) and his theorem from high school geometry. Not only did he come up with the Pythagorean theorem—the sum of the squares of the sides of a right triangle is equal to the square of the hypotenuse—but he also taught that Earth was a globe (not a cylinder, as Anaximander thought), fixed within a sphere that held the stars.

Anaxagoras Explains Eclipses and Aristarchus Puts the Sun in Its Place

Anaxagoras (ca. 500–ca. 428 B.C.E.) believed that the Sun was a large, red-hot body and that the Moon was much like Earth, complete with mountains and ravines. Most important, he theorized that solar eclipses were caused not by a dragon eating the Sun, but by the passage of the Moon between the Sun and Earth.

Next, Aristarchus of Samos (ca. 310–230 B.C.E.) proposed that Earth is not at the center of the universe (as others proposed), but that it orbits the Sun while also rotating. This early *heliocentric* theory, however, was rejected, and the Greeks as well as almost everyone who followed them held on to the *geocentric*—Earth-centered—model for millennia.

Star Words _____

Geocentric means Earth-centered, and the geocentric model of the universe is one in which Earth is believed to be at the center of the universe. **Heliocentric** means Sun-centered, which accurately describes our solar system, in which the planets and other bodies orbit the Sun.

Aristotle Lays Down the Law, and Ptolemy Paints a Picture

Fortunately for the hard-pressed astronomers of yore, total eclipses of the Sun are relatively rare events. More immediately, they had *nightly* occurrences they couldn't explain.

They could watch the sky all night and see the stars glide predictably across the sky, as if affixed to a moving celestial sphere. Likewise, the Moon made its traversal with perfect regularity. But there were (so far as the ancients could see) five heavenly bodies that didn't behave with this regularity. Mercury, Venus, Mars, Jupiter, and Saturn looked pretty much like stars, but, in contrast to stars, they wandered across the sky (as noted in Chapter 1, the Greeks called these celestial bodies *planetes*, or "wanderers").

Although the planets always remain near the Sun's apparent course in the sky, called the ecliptic, they have strange motions. Relative to the stars, planets seem to slow down and speed up, moving (from an earthbound observer's perspective) generally eastward (*prograde* motion), but sometimes westward (*retrograde* motion) with respect to the background stars.

How did the ancients explain this strange observation?

The Greek philosopher Aristotle (384–322 B.C.E.) formulated a picture of the solar system that put Earth at its center with all the other heavenly bodies orbiting around it. The orbits were described as perfect circles. However, Aristotle's simple geocentric (Earth-centered) model did not explain why the planets varied in brightness, why they moved at varying speeds, and why their motion was sometimes prograde and sometimes retrograde.

Generations of European astronomers wrestled with Aristotle's model in an effort to show how it was actually correct. The most impressive of these wrestling matches was fought by Claudius Ptolemaeus, a Greek astronomer better known as Ptolemy, who was active in Alexandria during C.E. 127–145. He developed the geocentric model into an impressively complex system of "deferents," large circular orbits centered on Earth, and "epicycles," small circles whose centers traveled around the circumferences of the deferents. Some 80 deferents and epicycles came into play, along with several other highly complex geometric arrangements, which enabled Ptolemy to account for many of the perplexing planetary motions he and others observed.

The cosmos according to Aristotle, as published in Peter Apian's Cosmographia *of 1524.*

(Image from Rice University)

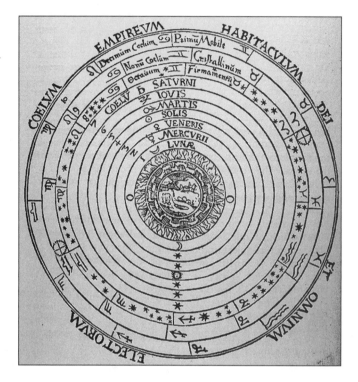

Close Encounter

Almost nothing is known about the life of Ptolemy, but his astronomical work was collected in a single great volume, at first called (in Greek) *He mathematike syntaxis* (*The Mathematical Collection*). The book was later known as *Ho megas astronomos* (*The Great Astronomer*), but perhaps is best known by what Arab astronomers called it in the ninth century, *Almagest,* essentially, *The Greatest Book.*

It is *Almagest* that has come down to us today. Divided into 13 books, it is nothing less than a synthesis of all of Greek astronomy, a star catalog started by Hipparchus and expanded by Ptolemy into an identification of 1,022 stars. In addition, *Almagest* presents Ptolemy's geocentric model of the solar system, which was generally accepted as an accurate predictive model until Copernicus offered a heliocentric alternative in 1543.

Astronomy Dies and Is Reborn

Even if today's astronomers knew nothing of the solar system, they would likely reject the Ptolemaic model on the grounds of its complexity. And for all of its intricacy, the Ptolemaic model was not a particularly accurate predictor of astronomical phenomena. A love of elegant simplicity also characterized the classical world, but Ptolemy's era was already falling away from classical elegance and toward the cobwebbed mysteries that so appealed to people in the Middle Ages, when complexity and obscurity, not simplicity and clarity, were taken as the hallmarks of truth.

Arabian Nights

The Middle Ages, or "Dark Ages," weren't dark everywhere, and astronomy didn't exactly wither and die after Ptolemy. Outside Europe, some exciting discoveries were being made.

In medieval Arabia, Abu 'Abd Allah Muhammad Ibn Jabir Ibn Sinan Al-battani Al-harrani As-sabi', more conveniently known as al-Battani (ca. 858–929), made important refinements to calculations of the length of the year and the seasons and demonstrated that the Sun's *apogee* (its farthest point from Earth) is variable. He also refined Ptolemy's astronomical calculations by replacing geometry with sleek trigonometry.

Another Arab, Al-Sûfi (903–986), meaningfully discussed the comparative brilliance of the stars, while other Arab astronomers contributed star maps and catalogues that were so influential that many of the star names in use today are of Arab origin (such as Betelgeuse, Aldebaran, and Algol), as are such basic astronomical terms as *azimuth* and *zenith*.

Heresy of a Polish Priest

In Europe, astronomy—as a truly observational science—did not revive until the Middle Ages had given way to the Renaissance. Nicolaus Copernicus was born in eastern Poland in 1473. A brilliant youth, he studied at the universities of Kraków, Bologna, Padua, and Ferrara, learning just about everything that was then known in the fields of mathematics, astronomy, medicine, and theology.

Copernicus earned great renown as an astronomer. Anxious to ensure that feast days and holy days were accurately observed, the Catholic Church asked him in 1514 for his opinion on how the current calendar might be reformed for greater precision. Copernicus responded that he could not give an opinion, because the positions of the Sun and Moon were not understood with sufficient accuracy.

Copernicus, as pictured in Charles F. Horne's Great Men and Famous Women (1894).

(Image from the authors' collection)

Copernicus questioned the Ptolemaic system, which was the basis of the calendar on which the Church relied. He read ancient Greek manuscripts and rediscovered the idea of a heliocentric (Sun-centered) universe. He concluded that putting the Sun at (or near) the center of a solar system with planets in orbit around it created a model that was "more pleasing to the mind" than what Ptolemy had proposed and medieval Europe accepted for so many centuries.

Copernicus wrote *De Hypothesibus Motuum Coelestium a Se Constitutis Commentariolus* (*A Commentary on the Theories of the Motions of Heavenly Objects from Their Arrangements*) in 1514, arguing that all of the motion we see in the heavens is the result of Earth's daily rotation on its axis and yearly revolution around the Sun, which is motionless at the center of the planetary system.

Astronomer's Notebook

Properly speaking, Copernicus's solar system was *heliostatic* (a motionless Sun) rather than *heliocentric* (a centered Sun); to account more accurately for planetary motion, he did not place the Sun at the precise center of the system, but offset it slightly. The perfection of shapes, and circles in particular, had been part of the astronomer's baggage since Greek times. It would take the brilliant and unconventional idea of elliptical, rather than circular, orbits—put forth by Johannes Kepler—to reconcile all the apparent irregularities of planetary motion.

Although Copernicus's purpose in reordering the planetary system might have been relatively modest—a method of making a better calendar—it soon became apparent that the new theory required a most profound revision of thought.

First: The universe had to be a much bigger place than previously imagined. The stars always appeared in the same positions with the same apparent brightness. But if Earth really were in orbit around the Sun, the stars should display a small but noticeable periodic change in position and brightness. Why didn't they? Copernicus said that the starry celestial sphere had to be so distant from Earth that changes simply could not be detected. Building on this explanation, others theorized an *infinite* universe, in which the stars were not arranged on a celestial sphere, but were scattered throughout space.

The second revision that a heliocentric universe required, while not as obvious, was even more basic.

Why do things fall? Aristotle explained that bodies fell toward their "natural place," which, he said, was the center of the universe. That explanation worked as long as Earth was considered to be the center of the universe. But now that it wasn't considered the center, how could the behavior of falling bodies be explained? The answers would have to wait until the late seventeenth century, when Isaac Newton published his *Principia*, including a theory of universal gravitation.

Profound as the astronomical and other scientific implications were, the emotional shock of the Copernican universe was even greater. Suddenly, Earth, with humankind upon it, was no longer the center of all creation, but was instead hurtling through space like a ball on a string. Why did we stay on the ball? What was the string? These were unanswered questions that must have been very unsettling for those who thought about them.

The Man with the Golden Nose

The astronomer Tycho Brahe (1546–1601) was born in Denmark shortly after the publication of Copernicus's *De Revolutionibus*. As a youth, he studied law, but was so impressed by astronomers' ability to predict a total solar eclipse on August 21, 1560, that he devoted himself to astronomy from that moment on and became a master of careful astronomical observation. Meticulous in science, he was reckless and flamboyant in his personal life, losing part of his nose in a duel. He replaced it with a glorious golden prosthesis.

Such was Tycho's renown that King Frederick II of Denmark gave him land and financed the construction of an observatory called Uraniborg (after Urania, the muse of astronomy). To this place journeyed some of the most respected scholars of the world, but when Frederick died in 1588, Tycho lost his patron, and he ultimately left Denmark, settling in Prague in 1599 as Imperial Mathematician of the Holy Roman Empire. There he was joined by a younger German astronomer, Johannes Kepler (1571–1630), who became his student and disciple.

Tycho Brahe in his observatory, as seen in an early print.

(Image from arttoday.com)

Kepler's Three Great Laws

Tycho Brahe died, ingloriously enough, in 1601, apparently from a burst bladder after drinking too much at a dinner party. Kepler inherited the master's mass of accumulated observational data and began trying to make theoretical sense of it all.

Johannes Kepler.

(Image from arttoday.com)

Tycho's data made clear to Kepler that the Copernican system of planets moving around the Sun in perfectly circular orbits wasn't correct. Although Kepler had predicted that with Tycho's data, he would solve the problem of planetary motion in a matter of days, it took him eight years of study, trial, and error before he finally concluded that the planets must orbit the Sun not in perfect circles, but in elliptical orbits (an *ellipse* is an oblong, or flattened, circle drawn around two foci instead of a single center point).

Kepler went on to state the fundamentals of planetary motion in three basic laws. That planets move in elliptical orbits, with the Sun at one focus of the ellipse, is known as Kepler's First Law. It resolved the discrepancies in observed planetary motion that both Ptolemy and Copernicus had failed to explain adequately.

Another apparent attribute of elliptical orbits determined Kepler's Second Law. It states that an imaginary line connecting the Sun to any planet would sweep out equal areas of the ellipse in equal intervals of time. (See the following "Close Encounter" sidebar for an explanation of focus and for a demonstration of Kepler's Second Law.) The second law explained the variation in speed with which planets travel. They will move faster when they are closer to the Sun. Kepler did not say *why* this was so, just that it was apparently so. Later minds would confront that *why*.

Close Encounter _____

The nature of an ellipse and Kepler's Second Law are both difficult to explain, but easy to visualize. You can draw an ellipse with the aid of a couple of tacks, string, a piece of cardboard, and a pencil. Push the two tacks into a piece of cardboard and loop the string around them (leave some play in the string). The tacks are the foci of the ellipse. The closer together the foci, the less elliptical (more perfectly circular) the ellipse. To draw an ellipse, simply place the pencil inside the loop of the string, and pull it taut–now, pull the pencil around the tacks all the way. Depending on the separation of the tacks, you will draw different looking ellipses.

To illustrate Kepler's Second Law, imagine the Sun as being located at the focus nearest *a* (that is, the tack nearest *a*). The pencil point is a planet, and the portion of the string connecting the pencil point to the tack nearest *a* is the imaginary line connecting the Sun to a planet.

To accurately model the orbit of the planet around the Sun, you would have to be able to draw the ellipse such that the pencil sweeps out equal areas in equal intervals of time. This wouldn't be an easy trick, because you would have to draw faster when the pencil/planet was closest to the tack/Sun (the position called *perihelion*) than when it was farthest from the Sun (*aphelion*). To make an area swept near *perihelion* equal to an area swept at *aphelion* you have to cover more of the circumference of the ellipse nearer the Sun-tack than you have to cover farther from the Sun-tack.

The first two of Kepler's laws were published in 1609. The third didn't appear until 10 years later and is slightly more complex. It states that the square of a planet's *orbital period* (the time needed to complete one orbit around the Sun) is proportional to the cube of its *semi-major axis* (half the longest distance across the ellipse, analogous to the radius of a circle). Because the planets' orbits, while elliptical, are very nearly circular, the semi-major axis can be considered to be a planet's average distance from the Sun. If the distance (a) is measured in *astronomical units* (AU), and the orbital period (P) is measured in years, Kepler's third law simply says that $P^2 = a^3$.

Star Words _____

An **astronomical unit** (A.U.) is equal to the mean distance between Earth and the Sun. It is a conveniently large unit for measuring the large distances in the solar system.

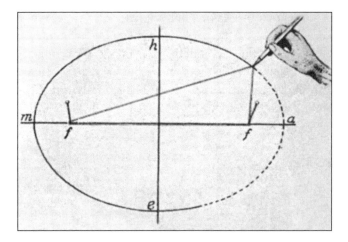

Drawing an ellipse to illustrate Kepler's Second Law. See the preceding Close Encounter sidebar for an explanation.

(Image from the authors' collection)

Galileo's Eye

While Kepler pored over Copernicus's data, an Italian astronomer, Galileo Galilei (1565–1642), directed his gaze skyward, amplifying his eyesight with the aid of a new Dutch invention (which we'll introduce in the next chapter), the telescope.

Through this instrument, Galileo explored the imperfect surface of the Moon, covered as it was with craters, "seas," and features that looked very much like the surface of Earth. Galileo was even more surprised to find that the surface of the Sun was blemished. These "sunspots" (which you'll investigate in Chapter 12), he noted, changed position from day to day. From this fact, Galileo didn't conclude that the spots changed, but that the Sun was rotating, making a complete revolution about once each month.

Galileo Galilei.

(Image from arttoday.com)

Galileo published his experimental proofs of a heliocentric solar system in 1610. Six years later, the Catholic Church judged the work heretical and banned it, as well as the work of Copernicus. Galileo defied the ban and, in 1632, published a comparison of the Ptolemaic and Copernican models, in which, of course, the Copernican view came out the winner. The next year, the Holy Inquisition forced Galileo to recant this "heresy" under threat of death, and placed him under house arrest for the rest of his life.

Close Encounter

Popular legend has it that Galileo dutifully repeated aloud what the Inquisitors ordered him to say, that Earth does not move, only to mutter under his breath, as he marched off to house arrest, "Still, it moves." House arrest didn't prevent Galileo from continuing to work or even from corresponding with other scientists. "Though my soul may set in darkness, it will rise in perfect light," he wrote. "I have loved the stars too fondly to be fearful of the night."

Newton's Brain

One of us knew a young man who owned a pocket watch, which he would habitually twirl, holding the end of the chain and allowing it to orbit around the focus of his finger. One day, the chain broke, sending the costly timepiece flying off into space—and against a wall, with predictably catastrophic results.

Why don't the planets suffer the same fate? For all he had accomplished, Kepler did not explain how the planets orbited the Sun without flying off into space, and *why* they traveled in ellipses.

The answers came in the late seventeenth century when an Englishman, Isaac Newton (1642–1727), one of the most brilliant theoreticians to ever live, formulated three laws of motion and the law of universal gravitation in a great work, *Philosophiae Naturalis Principia Mathematica* (*Mathematical Principles of Natural Philosophy*), better known simply as *Principia*.

Three Laws of Motion

Newton's First Law of Motion states that, unless acted upon by some external force, a body at rest remains at rest and a moving object continues to move forever in a straight line and at a constant speed. This property is known as *inertia*. The measure of an object's inertia is its *mass* (in effect, the amount of matter the object contains). The more massive an object, the greater its inertia.

Star Words

Inertia is the tendency of a body in motion to remain in motion in a straight line and at a constant velocity unless acted on by an external force; it is also the tendency of a body at rest to remain at rest unless acted upon by an external force. **Mass** can be thought of as the amount of matter an object contains or as a measure of an object's inertia.

This law explains why the planets move in nearly circular orbits—essentially because an external force (gravity) acts on each planet. Without gravity, the planets would all fly off in straight lines, like so many pocket watches.

Newton's Second Law of Motion states that the acceleration of an object is directly proportional to the force applied to the object and inversely proportional to the mass of the object. Pull two objects with the same force, and the more massive object will accelerate more slowly than the less massive one. Newton's Third Law of Motion states that forces don't act in isolation. If object A exerts a force on object B, object B exerts an *equal but opposite* force on object A. A hammer, for example, exerts a force on the nail, driving it into the wall. The nail exerts an equal and opposite force on the hammer, arresting its motion.

Weighty Matters

Throw a ball up into the air, and you will observe that it travels in a familiar curved (parabolic) trajectory: first up, up, up, leveling off, then down, down, down. Common sense tells us that the force of gravity pulls the ball back to Earth.

Newton's brilliance was in postulating not only that there is a force, gravity, that pulls the ball (or apple) back to Earth, but that such a force applies to everything in the universe that has mass. The gravitational force due to the mass of Earth also pulls on the Moon, holding it in its orbit, and pulls on each of us, keeping us in contact with the ground. Finally there was an answer to those who thought Earth could not be spinning and orbiting the Sun. What was there to keep us firmly footed on the earth? Newton had the answer: the force of gravity.

Sir Isaac Newton.

(Image from arttoday.com)

It's Not Just a Good Idea ...

In subsequent chapters discussing the planets, we will look more precisely at how mathematics describes planetary motion. For now, consider the following general points.

In *Principia*, Newton proposed that the force of gravity exerted by objects upon one another is proportional to the mass of the two objects, and weakens as the square of the distance between those objects increases.

Specifically, he postulated that the gravitational force between two objects is directly proportional to the product of their masses (mass of object A times mass of object B). So two objects, one very massive and the other with very little mass, will "feel" the same mutual attraction. In addition, he claimed that the force between two objects will decrease in proportion to the square of the distance. This "inverse-square law" means that the force of gravity mutually exerted by two objects, say 10 units of distance apart, is 100 times (10^2) weaker than that exerted by objects only 1 unit apart—yet this force never reaches zero. The most distant galaxies in the universe exert a gravitational tug on one another. These relations between mass, distance, and force comprise what is called Newton's Law of Universal Gravitation.

Consider the solar system, with the planets moving in elliptical orbits around the Sun. Newton's *Principia* explained not only what holds the planets in their elliptical orbits (an "inverse-square" force called gravity), but also predicted that the planets themselves (massive Jupiter in particular) would have a small but measurable effect on each other's orbits.

Close Encounter _____

Newton's mathematical prowess was legendary, and he had both the brilliance and the determination to solve problems that had stymied others for decades. His friend, the great astronomer Edmund Halley, and others had been thinking for some time about a way to account for the observed elliptical orbits of planets, and thought that if the force of gravity decreased with the square of the distance, that the elliptical orbits of Kepler could be explained. But Halley was unable to prove his hunch mathematically. One day, Halley asked Newton what would be the trajectory of a planet moving in a gravitational force field that decreased in strength with the square of the distance from the Sun. Newton instantly responded: "An ellipse."

Halley asked to see the mathematical proof, and a few months later, Newton sent him a letter in which he derived Kepler's Three Laws of Planetary Motion from the basis of a gravitational attraction that decreased with the square of the distance between the Sun and the planets. With Halley's encouragement, Newton began his masterwork, *Principia,* in which he put forth his famous Laws of Motion and the Law of Universal Gravitation.

For the first time, a scientist had claimed that the rules of motion on Earth were no different from the rules of motion in the heavens. The Moon was just a big apple, much farther away, falling eternally to Earth in its own way. The planets orbit the Sun following the same rules as a baseball thrown up into the air, and the pocket watch of Earth is held in its orbit by a chain called gravity.

The Least You Need to Know

◆ Ptolemy made a highly successful—and complicated—attempt to account for all the complexities of movement among the celestial bodies. His geocentric model of the solar system was accepted until better observations of planetary positions in the sixteenth and seventeenth centuries made it clear that the model had to be significantly revised.

◆ Seeking an explanation of the motion of the planets accurate enough to enable him to revise inaccurate European calendars, Copernicus put the Sun at the center of the solar system, with the planets, including Earth, in orbit around it.

◆ Tycho Brahe made extraordinary astronomical observations, but it was his student, Johannes Kepler, who reduced these data to three basic laws of planetary motion. These laws, based on elliptical planetary orbits, brought the heliocentric model of the solar system into close agreement with observations.

- Galileo used the newly invented telescope to provide experimental proofs of the Copernican idea.

- With the Sun-centered universe in place, Isaac Newton put forth in *Principia* the Three Laws of Motion and the Law of Universal Gravitation, which accounted for the forces behind planetary motion. For the first time, someone proposed that the laws governing the motion of objects here on Earth and in the heavens were one and the same.

Collecting Light

In This Chapter

- ◆ Overcoming the tyranny of vast distance in the universe
- ◆ Light as energy that conveys information
- ◆ A brief look at the spectrum and waves
- ◆ An introduction to the telescope
- ◆ Resolution and atmospheric interference
- ◆ The *Hubble Space Telescope* and other cutting-edge projects
- ◆ Astronomy: getting in on the action

You have every reason and right to look at the night sky as the greatest free show in the universe. Most amateur stargazers—in fact, most professional astronomers—are strongly attracted by the great beauty of the sky. But the sky is more than beautiful. Celestial objects are full of information just waiting to be interpreted—information like how distant the stars and galaxies are, how large they are, and whether they are moving toward or away from us.

How does the information reach us?

It travels to us in the form of electromagnetic radiation, a small fraction of which is visible light. This chapter begins by defining light and then covers how we can most effectively collect it to see more of the solar system and the universe.

Slices of Light

The universe is ruled by the tyranny of distance. That is, the universe is so vast that we are able to see many things that we will never be able to visit. Light is able to travel at extraordinary speeds (about 984,000,000 feet—300,000,000 meters—every second), but the light that we now see from many objects in the sky left those sources thousands, millions, or even billions of years ago. It is possible, for example, to see the Andromeda galaxy (millions of *light-years* away), even with the naked eye, but it's unlikely humans will ever travel there.

Star Words

A **light-year** is the distance light travels in one year—about 6 trillion miles (9,461,000,000,000,000 meters). In the vastness of space beyond the solar system, astronomers use the light-year as a basic unit of distance.

Well, Andromeda is about two million light-years away, and a *light-year* is the distance light travels in one year—some 6 trillion miles. Now, light can travel that far every year, so to get the distance to Andromeda, you multiply the velocity of light (6 trillion miles in a year) by the amount of time it took the light to get here (2 million years), and you get a lot of miles—approximately a 1 with 19 zeroes after it. Here's another way to think about these unbelievable distances: If you could travel at the speed of light (an impossibility, according to Einstein's theory of relativity, as you will see in later chapters), it would *still* take you two million years to reach Andromeda.

But we can't travel at anywhere near the speed of light. Right now, the fastest rockets are capable of doing 30,000 miles per hour (48,000 km/h). Maybe—someday—technology will enable us at least to approach the speed of light, but that still means a trip of two million years up and two million back. To put that into perspective, all of recorded history consumes no more than 5,500 years, and civilization, perhaps 10,000 years.

Why not go faster than the speed of light? In Chapter 4, you'll see that, according to our understanding of space and time, the speed of light is an absolute speed limit that cannot be exceeded.

So revel in the fact that, on a clear night, you are able to gaze at the Andromeda galaxy, an object so distant that no human being will likely ever visit it.

Spaceships might be severely limited as to how fast they can travel, but as we've said, the information conveyed by *electromagnetic radiation* can travel at the speed of light. The information from Andromeda, it is true, isn't exactly fresh news by the time we get it. In fact, the photons that we are receiving from Andromeda left that galaxy long before *Homo sapiens* walked the earth. But everything we know about Andromeda and almost all other celestial bodies (aside from the few solar system objects we have visited with probes or landers) we know by analyzing their electromagnetic radiation.

Electromagnetic radiation travels through the vacuum of space in waves, which we shall also examine in some detail in Chapter 4. A wave—think of a water wave—is not a physical object, but a pattern of up-and-down or back-and-forth motion created by a disturbance. Waves are familiar to anyone who has thrown a rock in a pond of still water or watched raindrops striking a puddle. The wave pattern in the water, a series of concentric circles, radiates from the source of the energy: the impact of the rock or the raindrop. If anything happens to be floating on the surface of the water—say a leaf—the waves will transfer some of the energy of the splash to the leaf and cause it to oscillate up and down. The important thing to remember about waves is that they convey both energy and information. Even if you didn't actually see the rock or the raindrop hit the water, you would be able to surmise from the action of the waves that something had disturbed the surface of the water at a particular point.

Star Words

Electromagnetic radiation transfers energy and information from one place to another, even in the vacuum of space. The energy is carried in the form of rapidly fluctuating electric and magnetic fields and includes visible light in addition to radio, infrared, ultra-violet, x-ray, and gamma-ray radiation.

The type of energy and information created and conveyed by electromagnetic radiation is more complex than that created and conveyed by the waves generated by a splash in the water. You will have to wait until Chapter 4 for a little lesson in wave anatomy, but do take a moment now to make sure that you understand two properties of waves: *wavelength* and *frequency*. Wavelength is the distance between two adjacent wave crests (high points) or troughs (low points), measured in meters. Water waves might have wavelengths of a few meters, radio waves of a few centimeters, and the wavelengths of optical light are very short. Frequency is the number of wave crests that pass a given point per unit of time (and has units of 1/second). By convention, frequency is measured in hertz (Hz). One hertz is equivalent to one crest-to-crest cycle per second and is named in honor of the nineteenth-century German physicist Heinrich Rudolf Hertz.

You might think of the light from your reading lamp as being very different from the x-rays your dentist uses to diagnose an ailing tooth, but both are types of electromagnetic waves, and the only difference between them is their wavelengths. Frequency and wavelength of a wave are inversely proportional to one another, meaning that if one of them gets bigger, the other one must get smaller. The particular wavelength produced by a given energy source (a star's photosphere—its visible outer layer—or a planetary atmosphere) determines whether the electromagnetic radiation produced by that source is detected at radio, infrared, visible, ultraviolet, x-ray, or gamma-ray wavelengths.

The waves that produce what we perceive as visible light have wavelengths of between 400 and 700 nanometers (a nanometer is 0.000000001 meter, or $1 - 10^{-9}$ m) and frequencies of somewhat less than 10^{15} Hz. Light waves, like the other forms of electromagnetic radiation, are produced by the change in the energy state of an atom or molecule. These waves, in turn, transmit energy from one place in the universe to another. The special nerves in the retinas of our eyes, the emulsion on photographic film, and the pixels of a CCD (Charge Coupled Device) electronic detector are all stimulated (energized) by the energy transmitted by waves of what we call visible light. That is why we "see."

The outer layers of a star, its photosphere, consist of extremely hot gas radiating some fraction of the huge amounts of energy that a star generates in its core through nuclear fusion (see Chapter 12 for more on this). That energy is emitted at some level in all portions of the *electromagnetic spectrum*, so that when you look at a distant or nearby star (the Sun) with your eyes, you are receiving a small portion of that energy.

Star Words

The **electromagnetic spectrum** is the complete range of electromagnetic radiation, from radio waves to gamma waves and everything in between, including visible light.

Buckets of Light

Of course, the fraction of the emitted energy you receive from a very distant star—or even a whole galaxy, like far-off Andromeda—is very small, having been diminished by the square of the distance (but never reaching zero). Imagine a sphere surrounding a distant star. As the sphere becomes larger and larger (that is, as you get farther and farther from the star), the same amount of energy passes through ever larger spheres. Your eye (or your telescope) can be thought of as a very tiny fraction of the sphere centered on that distant star. You are collecting as much light from the distant source as falls into your "light bucket." If your eye is a tiny "bucket," then a 4-inch amateur telescope is a slightly larger bucket, and the *Hubble Space Telescope* is an even larger one. The larger the bucket, the more light you can "collect." And if you collect more light in your bucket, you get more information.

One early question among astronomers (and others) was, "How can we build a better bucket than the two little ones we have in our head?"

The answer came in the early seventeenth century.

The Telescope Is Born

In 1608, lens makers in the Netherlands discovered that if they mounted one lens at either end of a tube and adjusted the distance between the lenses, the lens that you put to your eye would magnify an image focused by the lens at the far end of the tube. In effect, the lens at the far end of the tube gathered and concentrated (focused) more light energy than the eye could do on its own. The lens near the eye enlarged to various degrees that concentrated image. This world-changing invention was dubbed a telescope. The word *telescope* comes from Greek roots meaning "far-seeing." Optical telescopes are arrangements of lenses and/or mirrors designed to gather visible light efficiently enough to enhance observation of distant objects and phenomena.

Many, perhaps most, inventions take time to gain acceptance. Typically, there is a lapse of more than a few years between the invention and its practical application. Not so with the telescope. By 1609, within a year after the first telescopes appeared, the Italian astronomer Galileo Galilei demonstrated their significance in military matters (seeing a distant naval foe), and was soon using them to explore the heavens. The largest of his instruments was quite small, with only modest magnifying power, but Galileo was able to use this tool to describe the valleys and mountains on the Moon, to observe the phases of Venus, and to identify the four largest moons of Jupiter.

Refraction ...

Galileo's instrument, like all of the earliest telescopes, was a *refracting telescope*, which uses a glass lens to focus the incoming light. For all practical purposes, astronomical objects are so far away from us that we can consider that light rays come to us parallel to one another—that is, unfocused. Refraction is the bending of these parallel rays. The convex (bowed outward) piece of glass we call a lens bends the incoming rays so that they all converge at a point called the *focus*, which is behind the lens directly along its axis. The distance from the cross-sectional center of the lens to the focus is called the *focal length* of the lens. Positioned behind the focus is the eyepiece lens, which magnifies the focused image for the viewer's eye.

Modern refracting telescopes consist of more than two simple lenses. At both ends of the telescope tube, compound (multiple) lenses are used, consisting of assemblies of individual lenses (called elements) designed to correct for various distortions simple lenses produce.

Star Words

A **refracting telescope**, or **refractor**, creates its image by refracting (bending) light rays in glass lenses. The point at which those incoming parallel bent rays converge is called the **focus**, and the distance from the cross-sectional center of the lens to the focus is the **focal length** of the lens.

Diagram of a refracting telescope. CF represents the objective lens and LL the eyepiece. The observer's eye is identified by E.

(Image from arttoday.com)

... or Reflection?

The refracting telescope was one of humankind's great inventions, rendered even greater by the presence of a genius like Galileo to use it. However, the limitations of the refracting telescope soon became apparent:

- ♦ Even the most exquisitely crafted lens produces distortion, the chief of which is chromatic aberration.

- ♦ Excellent lenses are expensive to produce because both sides of a lens have to be precision crafted and polished. Mirrored surfaces require polishing only one side.

- ♦ Generally, the larger the lens, the greater the magnification and the brighter the image; however, large lenses get heavier faster than large mirrors. Lenses have volume, and the potential for imperfections (such as bubbles in the glass) is higher in a large lens. All of this means that large lenses are much more difficult and expensive to produce than small ones.

Recognizing the deficiencies of the refracting telescope, Isaac Newton developed a new design, the reflecting telescope, in 1668.

Diagram of a Newtonian reflector. Light enters at the left and is focused by the primary mirror (M) at the back of the telescope. The focused image is sent by a secondary mirror (G) through the eyepiece (LL). The observer's eye is labeled E.

(Image from arttoday.com)

Instead of the convex lens of a refractor, the reflector uses a concave mirror (shaped like a shallow bowl) to gather, reflect, and focus incoming light. The hollow side of your breakfast spoon is essentially a concave mirror (the other side is a convex one).

This curvature means that the focal point is in front of the mirror—between the mirror and the object being viewed. Newton recognized that this was at best inconvenient—your own head could block what you are looking at—so he introduced a secondary mirror to deflect the light path at a 90-degree angle to an eyepiece mounted on the side of the telescope.

Astro Byte

Newton usually gets sole credit for inventing the reflecting telescope, but another Englishman, John Gregory, actually beat him to it, with a design created in 1663. It was, however, the Newtonian reflector that caught on. The French lens maker Guillaume Cassegrain introduced another variation on the reflector design in 1672. In his design, there is a primary and a secondary mirror, and the focal point of the primary mirror is located behind the primary mirror surface, requiring a hole in the center of the primary mirror.

Refracting telescope design continued to develop, culminating in the 40-inch (that's the diameter of the principal lens) instrument at Yerkes Observatory in Williams Bay, Wisconsin, installed in 1897. But due to the limitations just mentioned, the biggest, most powerful telescopes have all been reflectors, such as the 100-inch reflector installed at the Mount Wilson Observatory (near Pasadena, California) early in the century.

The 100-inch reflector at Mount Wilson Observatory.

(Image courtesy of Carnegie Institution of Washington)

Variations on an Optical Theme

Although the two major types of optical telescopes are the refractor and the reflector, there are many variations in reflector design. Some larger reflecting telescopes employ a Cassegrain focus, in which the image from the primary mirror is reflected to a secondary mirror, which again reflects the light rays down through an aperture (hole) in the primary mirror to an eyepiece at the back of the telescope instead of at the side of the telescope.

A coudé-focus (*coudé* is French for "bent") reflector sends light rays from the primary mirror to a secondary mirror, much like a Cassegrain. However, instead of focusing the light behind the primary mirror, another mirror is employed to direct the light away from the telescope, through an aperture, and into a separate room, called the coudé-focus room. Here astronomers can house special imaging equipment that might be too heavy or cumbersome to actually mount to the barrel of the telescope.

Reflecting telescopes have their problems as well. The presence of a secondary mirror (or a detector, in the case of a prime-focus reflector) means that some fraction of the incoming light is necessarily blocked. Also, the spherical shape of the reflector introduces spherical aberration, light being focused at different distances when reflecting from a spherical mirror. If not corrected, this aberration produces blurred images. One common solution to spherical aberration is to use a very thin "correcting" lens at the top of the telescope. This type of telescope is called a Schmidt-Cassegrain, and is a popular design for some high-end amateur telescopes.

Size Matters

In 1948, the Hale telescope at Mount Palomar, California, was dedicated. Its 200-inch (5-meter) mirror was the largest in the world until 1974, when the Soviets completed a 74-ton, 236-inch (6-meter) mirror, which was installed at the Special Astrophysical Observatory in Zelenchukskaya in the Caucasus Mountains.

Astro Byte

In theory, the 6-meter Russian telescope in the Caucasus can detect the light from a single candle at a distance of 14,400 miles. The presence of Earth's atmosphere and other real-world factors don't permit the practical achievement of this theoretical potential, however.

In 1992, the first of two Keck telescopes, operated jointly by the California Institute of Technology and the University of California, became operational at Mauna Kea, Hawaii. The second Keck telescope was completed in 1996. Each of these instruments combines 36 71-inch (1.8-meter) mirrors into the equivalent of a 393-inch (10-meter) reflector. Not only do these telescopes now have the distinction of being the largest telescopes on Earth, they are also among the highest (of those based on Earth), nestled on an extinct volcano 2.4 miles above sea level.

Astronomer's Notebook

Astronomers speak of the angular resolution of a telescope, which is a measurement of the smallest angle separating two objects that are resolvable as two objects. Generally, Earth's major optical and infrared telescopes, located at the best sites, can resolve objects separated by as little as 1" (that is, 1 arcsecond, which is $\frac{1}{60}$ of 1 arcminute, which, in turn, is $\frac{1}{60}$ of 1 degree). The theoretical resolution of these telescopes is much higher than this value, but turbulence in Earth's atmosphere means that, except for exceptional nights, this is the best that an Earth-based optical telescope can do. No matter how big the telescope, conventional telescopes cannot have resolutions higher than this value unless they employ adaptive optics. In recent years, the use of "adaptive optics," which uses computer technology to make real-time adjustments in telescope optics to compensate for atmospheric turbulence, has improved the resolution to the subarcsecond level.

The Power to Gather Light

Why this passion for size?

As we mentioned before, the bigger the bucket, the more light you can collect, so the more information you can gather. The observed brightness of an object is directly proportional to the area (yes, area; not diameter) of the primary mirror. Thus, a 78-inch (2-meter) diameter mirror yields an image 4 *times* brighter than a 39-inch (1-meter) mirror, because area is proportional to diameter squared, and the square of 2 (2 times 2) is 4. A 197-inch (5-meter) mirror would yield images 25 *times* brighter (5 times 5) than a 1-meter mirror, and a 393-inch (10-meter) mirror would yield an image 100 times brighter than a 1-meter mirror.

Now, things that are farther away are always going to be more faint. It should be obvious that a 100-watt light bulb will appear more faint if it is 1 mile away versus 1 foot away. Thus, a telescope that can see more faint objects is able to see things that are farther away. So, in general, the bigger the telescope, the more distant are the objects that can be viewed.

The Power to Resolve an Image

Collecting more light is only one advantage of a large telescope. Such instruments also have greater *resolving power*—that is, the ability to form distinct and separate images of objects that are close together. Low resolution produces a blurred image; high resolution produces a sharp image.

Star Words

Resolving power is the ability of a telescope to render distinct, individual images of objects that are close together.

Twinkle, Twinkle

Theoretically, the giant Hale telescope at Mount Palomar is capable of a spectacular angular resolution of .02" (or 20 milliarcseconds); that would be its resolution in the absence of complicating factors like Earth's atmosphere. In actual practice, it has a resolution of about 1" (1 arcsecond). The source of this limit is related to the reason that stars twinkle. Earth's turbulent atmosphere stands between the telescope's gigantic primary mirror and the stars, smearing the image just as it sometimes causes starlight viewed with the naked eye to shimmer and twinkle. If you took a still photograph of a twinkling star through a large telescope, you would see not a pinpoint image, but one that had been smeared over a minute circle of about 1 arcsecond. This smeary circle is called the seeing disk, and astronomers call the effect of atmospheric turbulence *seeing*. When weather fronts are moving in (even if the skies appear clear), or have just moved out, the seeing can be particularly bad.

> **Star Words**
>
> **Seeing** is a measure of the sharpness of telescopic images as a result of atmospheric turbulence. "Good seeing" means conditions that are minimally impacted by the effect of turbulence.

Computer Assistance

Beginning in the late nineteenth century, most serious telescope viewing was done photographically, not in real time, but by studying photographic plates exposed at the focus of a telescope. Photographic methods allowed astronomers to make longer observations, seeing many more faint details than could ever be distinguished with visual observing. In recent years, chemical-based photography has yielded to digital photography, which records images not on film but on CCDs (charge-coupled devices)—in principle the same device at the focal plane of your camcorder lens or digital camera.

> **Close Encounter**
>
> This might be a good time to take a break and turn to your computer. Log on to the World Wide Web and point your browser to the site of the Space Telescope Science Institute at www.stsci.edu and peruse some of the wonderful images from the *Hubble Space Telescope* (which we will discuss in just a moment).

CCDs are much more sensitive than photographic film, which means they can record fainter objects in briefer exposure times; moreover, the image produced is digital and can be directly transferred to a computer. Remember the sound of old-fashioned 12-inch, vinyl LP records? Even the best of them had a hiss audible during quiet musical passages. CDs, recorded digitally, changed all that by electronically filtering out the nonmusical noise found at high frequencies. Analogous digital computer techniques can be used to filter out the "visual noise" in an image to improve its quality.

Fun House Mirrors

Despite atmospheric turbulence, recent technology has made it possible to break the 1-arcsecond barrier.

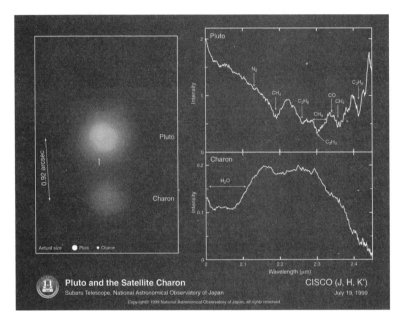

An image of the planet Pluto and its companion Charon taken with the Subaru 8.3-m telescope resolves the two closely orbiting bodies into separate points of light. The apparent separation of Pluto and Charon is only about 0.9", beyond detection without adaptive optics.

Adaptive optics systems enable astronomers to correct the distortions introduced by the atmosphere with distortions of their own. The distortions are made to another reflective surface inserted into the optical path, the path that light follows through the telescope. The idea is that if the distortions can be removed quickly enough (in real time), then large telescopes would have both of the advantages that they should have, namely more sensitivity and more resolution. This technology is very dependent on fast computers and rapidly movable motors that can make tiny, precise adjustments to the surfaces of small mirrors. However, as the technology is perfected, ground-based telescopes will be able to make images as sharp as those made from space.

An Observatory in Space: The *Hubble Space Telescope*

Another way to escape the bad seeing caused by Earth's atmosphere is to get above the atmosphere. That is just what NASA, in conjunction with the European Space Agency, did with the *Hubble Space Telescope* (HST). High above Earth's atmosphere, the HST regularly achieves its theoretical resolution of about less than 0.05" (50 milliarcseconds).

The HST was deployed from the cargo bay of the space shuttle *Discovery* in 1990. The telescope is equipped with a 94-inch (2.4-meter) reflecting telescope, capable of 10 times the angular resolution of the best Earth-based telescopes and approximately 30 times more sensitive to light, not because it is bigger than telescopes on Earth, but because it is above Earth's atmosphere. Unfortunately, due to a manufacturing flaw, the curvature of the 2.4-meter mirror was off by literally less than a hair (it was too flat by ⅕₀ of the width of a human hair), which changed its focal length. The telescope still focused light, but not where it needed to, in the plane of the various detectors. Astronauts aboard the shuttle *Endeavour* rendezvoused with the HST in space in 1993 and made repairs—primarily installing a system of small corrective mirrors. HST then began to transmit the spectacular images that scientists had hoped for and the world marveled at.

The HST is scheduled to be replaced by the Next Generation Space Telescope (NGST) by the year 2010.

Astro Byte

At a cost of $3 billion, the *Hubble Space Telescope* is one of the most expensive scientific instruments ever made. However, considering the wealth of scientific information it has produced, the *Hubble* is a bargain.

Close Encounter

As we have said, yet another way to achieve better resolution is to observe at another wavelength. The VLA or Very Large Array was dedicated in 1980, and has been operating continuously since then. It is located near the town of Socorro, New Mexico, and consists of 27 25-meter antennas that are connected to form an *interferometer*, a coordinated array of receiving antennas, that acts as a single telescope. In the largest of its variable configurations and at the highest frequency, the VLA has a spatial resolution of almost 0.03", or three one hundredths of an arcsecond. That resolution is about the same as the HST—even though the VLA is an Earth-based telescope. How can this be? We'll talk more about radio astronomy in Chapter 5.

Participatory Astronomy

At nearly $3 billion for the *Hubble Space Telescope*, astronomy can be a dauntingly expensive pursuit. Fortunately, you don't have to spend quite that much to get started. In fact, you don't have to spend anything. A lot of observation can be done with the naked eye, and many local communities have active amateur astronomers who would be happy to let you gaze at the heavens through their telescopes. Investing in a good pair of binoculars is also a good way to start your sky viewing.

Buying—or making—and using your own telescope can be rewarding and even thrilling. It can also be disappointing. For example, you won't be very happy if you can never get away from the light pollution of the city or suburbs. And a cheaply made telescope will frustrate you with dark and blurry images, whereas a more costly device might send you into the black hole of credit card debt or cause a different kind of frustration as you try to deal with an instrument more complicated and temperamental than you really want to deal with.

Take time to learn about amateur telescopes, including what features to look for, how much to spend, and what to do with the thing once you've bought it. Check out any of these books: *Backyard Astronomy: Your Guide to Starhopping and Exploring the Universe*, edited by Robert Burnham (Time Life, 2001); *Turn Left at Orion*, by Guy J. Consolmagno (Cambridge University Press, 1995); and *The Backyard Astronomer's Guide*, by Terence Dickinson and Alan Dyer (Firefly, 2002).

And remember, you can get deeply involved in astronomy without owning or even using a telescope. If you have a personal computer (you do, don't you?), consider purchasing an astronomy software package, such as *The Sky* (from Bisque Software) or *Starry Night* ("Pro" or "Backyard" versions, from Space Software). Don't want to buy software in shrink wrap? Take a look at "CVC's Astronomy Freeware and Software" at www.cvc.org/astronomy/freeware.htm and Astronomy—Free Software at http://freeware.intrastar.net/astronmy.htm for a rundown of popular packages, including many that are freely downloadable.

But you don't need to buy or even download free software if you don't want to. Try surfing the Internet into outer space. A great place to start is www.nasa.gov, which has fantastic images of space exploration. *Hubble Space Telescope* images are available at http://hubble.nasa.gov and elsewhere on the Internet, and images from the Very Large Array and the Greenbank Telescope are available at www.nrao.edu. Another great source is the Jet Propulsion Laboratories web site at www.jpl.nasa.gov. And these are just for starters. Any good search engine will find thrilling images, including many from ongoing space missions as well as earthbound observatories.

Astronomy can become a life's work or something you do for pleasure after your day job. It can be an expensive hobby (there are plenty of dedicated amateurs who take out home equity loans to build backyard observatories around telescopes and ancillary equipment costing some tens of thousands of dollars), or it can be pretty close to free of charge. Take your pick.

The Least You Need to Know

- ◆ Light is a form of electromagnetic radiation. Radiation carries energy and conveys information.

- ◆ Objects in space produce or reflect the various forms of electromagnetic radiation (including radio, infrared, visible light, ultraviolet, x-rays, and gamma rays); this radiation is what we see with our eyes or detect with special instruments.

- ◆ The two basic optical telescope types are the lens-based refractors and the mirror-based reflectors.

- ◆ The two main functions of telescopes are to collect light and resolve objects. Larger telescopes (barring the effects of Earth's atmosphere) are better able to perform both these functions.

- ◆ New technologies, such as adaptive optics, allow ground-based telescopes to achieve much sharper images while maintaining the convenience and lower cost of being on the ground.

Chapter 4

Over the Rainbow

In This Chapter

- ◆ Understanding electromagnetic radiation
- ◆ Transmission of energy by waves
- ◆ How light travels through space
- ◆ A tour of the electromagnetic spectrum
- ◆ Our atmosphere as ceiling and skylight
- ◆ Using the electromagnetic spectrum to get information from the sky

The light we receive from distant sources is generated on the tiniest of scales. To explore the largest objects, such as galaxies, we first have to understand the smallest of objects: atoms and the particles that make up atoms. The photons that we detect with our eyes and catch with our telescopes were generated in many different ways: sometimes by electrons hopping between different orbital levels in an atom, or other times by the energetic collisions of atomic nuclei. In this chapter, we explore the ways in which photons of light arise, how they get from there to here, and what they can tell us about the objects that we observe.

We have concentrated thus far on optical photons (the ones that you can see with your eyes). As it turns out, our eyes respond to "visible" wavelengths because that is where the peak of the emission from the Sun is located in the electromagnetic spectrum. If your eyes were more sensitive to infrared radiation, for example, you would see some things you can't now see (body heat), but you'd miss a lot of other useful stuff (such as the visual appearance of that body that's giving off heat).

In this chapter, we're going to talk more about visible light and the electromagnetic spectrum, of which visible light is a tiny subset. Think of it this way: If the electromagnetic spectrum is represented by a piano keyboard, then the visible part of the spectrum is but a single key or note. In the cosmic concerto, there are many notes, and we want to be able to hear them all. Remember this astounding fact from Chapter 3: When you look at the light from the Sun or a distant star, you are witnessing the product of nuclear fusion reactions that are, every second, releasing more energy than any atomic explosion Earth has ever witnessed. Yet it is not just brute energy, but also information from the sky. Let's take a closer look.

Making Waves

Electromagnetic radiation sounds like dangerous stuff—and, in fact, some of it is. But the word *radiation* need not set off air raid sirens in your head. It just describes any way energy is transmitted from one place to another without the need for a direct physical connection between the two places. In this book, we use it as a general term to describe any form of light. It is important that radiation can travel without any physical connection, because space is essentially a vacuum; that is, much of it is empty (at least as far as any scale meaningful to human beings goes). If you went on a space walk clicking a pair of castanets, no one would hear your little concert. Sound is transmitted in waves, but not as radiation. Sound waves require some medium to travel in. So despite what most surround-sound science fiction movies would lead you to believe, explosions in space are silent. Light (and other forms of electromagnetic radiation) requires no such medium to travel (although many physicists tried in vain to detect a medium, which they called the ether; we'll talk more about this fact in a moment).

The electromagnetic part of the phrase denotes the fact that the energy is conveyed in the form of fluctuating electric and magnetic fields. These fields require no medium to support or sustain them.

Anatomy of a Wave

We can understand how electromagnetic radiation is transmitted through space if we appreciate that it involves waves. What is a wave? The first image that probably jumps to mind is that of ocean waves. And ocean waves do have some aspects in common with the kind of waves that are used to describe electromagnetic radiation. One way to think of a wave is that it is a way for energy to be transmitted from one place to another without any physical matter being moved from place to place. Or you might think of a wave as a disturbance that carries energy and that occurs in a distinctive and repeating pattern. A rowboat out in the ocean will move up and down in a regular way as waves pass it. The waves do transmit energy to the shore (think of beach erosion), but the rowboat will stay put.

That regular up-and-down motion that the rowboat experiences is called simple harmonic motion. But there are two important differences with electromagnetic radiation: The sources of waves are things on atomic scales (electrons and the nuclei of atoms), and no medium is required for electromagnetic waves to travel through space. The "pond" of space consists only of electric and magnetic fields, and photons of light are ripples in that ghostly pond.

Astronomer's Notebook

Recall from Chapter 3 that wave frequency is expressed in a unit of wave cycles per second, called the hertz, abbreviated Hz. Wavelength and frequency are inversely related; that is, if you double the wavelength, you automatically halve the frequency, and if you double the frequency, you automatically halve the wavelength. Multiply wavelength by frequency, and you get the wave's velocity. For electromagnetic radia-

---

Waves come in various shapes, but they all have a common anatomy. They have *crests* and *troughs*, which are, respectively, the high points above and low points below the level of an undisturbed state (for example, calm water). The distance from crest to crest (or trough to trough) is called the *wavelength* of the wave. The height of the wave—that is, the distance from the level of the undisturbed state to the crest of the wave—is its *amplitude*. The amount of time it takes for a wave to repeat itself at any point in space is its *period*. On our pond, the period is the time between the passage of wave crests as seen by an observer in the bobbing row boat. The number of wave crests that pass a given point during a given unit of time is called the *frequency* of the wave. If many crests pass a point in a short period of time, we have a high-frequency wave. If few pass that point in the same amount of time, we have a low-frequency wave. The frequency and wavelength of a wave are inversely proportional to one another, meaning that as one gets bigger, the other gets smaller. *High*-frequency radiation has *short* wavelengths.

The parts of a wave.

(Image from the authors' collection)

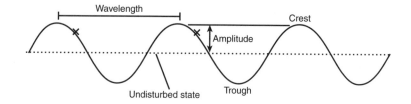

New Wave

If you don't happen to like math, don't panic. Just visualize pebble-generated waves rippling across a pond, and you'll understand the basic concept of waves.

But wait a minute. There is something wrong with our ripples in a pond as a model of electromagnetic radiation.

Water is a medium, a substance, something through which waves are transmitted. Space, we have said, is very nearly a vacuum, almost nothing. How, then, do waves move through it?

Astro Byte

Many people find it difficult to accept the idea of forces acting without any physical connection between them. Some scientists accused Newton of absurdity and even madness for having set forth a theory of gravity that proposed such action.

This is a question that vexed physicists for centuries. Sound, a wave, could not travel through a vacuum, so why could light, which is also a wave?

At first, most scientists refused to believe that space could be empty. They knew it didn't have air, as on Earth, but they suggested it was filled with another substance, which they called "the ether." But a series of experiments in the late nineteenth century made it clear that ether didn't exist and that although light could be studied as a wave, it was a very different kind of wave than, say, sound.

Big News from Little Places

The Greek philosopher Democritus (ca. 460–ca. 370 b.c.e.) was partially right: Matter does consist of atoms. But he would have been fascinated to know that the story doesn't end there. Atoms can be further broken down into electrons, protons, and neutrons, and the latter two are made of even smaller things called quarks. Electrons carry a negative electric charge, and protons a positive charge. Neutrons have a mass almost equal to a proton, but as their name implies, neutrons are neutral, with no positive or negative charge. All charged particles (like protons and electrons) are surrounded by what we call an electric field; those in motion produce electromagnetic radiation.

James Clerk Maxwell (1831–1879) first explored what would happen if such a charged particle were to oscillate, or move quickly back and forth. He showed that a moving charged particle created a disturbance that traveled through space—*without the need for any medium!* Particles in space are getting banged around all the time. Atoms collide, electrons are accelerated by magnetic fields, and each time they move, they pull their fields along with them, sending electromagnetic "ripples" out into space.

In short, information about the particle's motion is transmitted through space by a changing electric and magnetic field. But a field is not a substance. It is a way in which forces can be transmitted over great distances without any physical connection between the two places. The force of gravity, which we have discussed, can also be thought of as a field.

Let's turn to a specific example: A star is made up of innumerable atoms, most of which are broken into innumerable charged particles due to unimaginably hot stellar temperatures. A star produces a great deal of energy (by nuclear fusion, which we'll discuss in Chapter 12). This energy causes particles to be in constant motion. In motion, the charged particles are the center points of electromagnetic waves (disturbances in the electromagnetic field) that move off in all directions. A small fraction of these waves reaches the surface of Earth, where they encounter other charged particles. Protons and electrons in your eyes, for instance, oscillate in response to the fluctuations in the electric field. As a result, you perceive light: an image of the star.

If you happened to have, say, the right kind of infrared-detecting equipment with you, electrons in that equipment would respond to a different wavelength of vibrations originating from the same star. Similarly, if you were equipped with sufficiently sensitive radio equipment, you might pick up a response to yet another set of proton and electron vibrations.

Remember, it is not that the star's electrons and protons have traveled to Earth, but that the waves they generated so far away have excited other electrons and protons here. Think of it as an interstellar handshake.

Astronomer's Notebook

All electromagnetic waves—whether visible light, invisible radio waves, x-rays, or gamma radiation—move (in a vacuum) at the speed of light, approximately 186,000 miles per second. That's fast, but it's hardly an infinite, unlimited speed. Remember the Andromeda galaxy from Chapter 3? We can see it, but the photons of light we just received from the galaxy are 2 million years old. Now, *that's* a long commute.

Full Spectrum

Often, when people get excited, they run around, jump up and down, and shout without making a whole lot of sense. But when atomic particles get excited, they can produce energy that is radiated at a variety of wavelengths. In contrast to the babble of an excited human throng, this electromagnetic radiation can tell you a lot, if you have the instruments to interpret it.

Your eyes, one such instrument, can interpret electromagnetic radiation in the 400 to 700 *nanometer* (or 4000 to 7000 *Angstrom*) wavelength range. A nanometer (abbreviated nm) is one billionth of a meter, or 10^{-9} meter. An Angstrom (abbreviated A) is 10 times smaller, or 10^{-10} meter. But that is only a small part of the spectrum. What about the rest?

The Long and the Short of It

The senior author of this book is a radio astronomer. When he tells people this, he is typically met with a blank stare. The second most common response is, "Oh, so you listen to the stars." This popular misconception is reinforced by images of Jodie Foster, star of the 1997 sci-fi blockbuster *Contact*, "listening" to the pulses of a pulsar. In fact, you *can* "hear" radio sources on your radio (we'll tell you how in the very next chapter), and when most people hear the word *radio*, they think of the box in their car that receives radio signals broadcast from local towers, amplifies them, then uses them to drive a speaker, producing sound waves heard with the ears. But radio waves themselves are as silent as optical light or x-rays or gamma rays.

Radio waves are simply a form of electromagnetic radiation that has very long wavelengths.

So the only difference between radio waves and visible light waves is the length of the wave (or the frequency, which is always inversely related to wavelength). Indeed, all of the forms of electromagnetic radiation represented across what we call the electromagnetic spectrum—radio waves, infrared, visible light, ultraviolet, x-rays, and gamma rays—are transmitted at the speed of light as waves differentiated only by their wavelength and frequency. Radio waves are at the low end of the spectrum, which means their waves are big (on the scale of millimeters to meters in size) and their frequency, therefore, low—in the mega- (million) to giga- (billion) Hertz range. Look at the dial face of a radio tuner. You should notice that the FM band is rated in MHz, and the AM band in kHz, or 1,000 (kilo-) hertz. Gamma rays, in contrast, are at the high end of the spectrum, with very short wavelengths and very high frequencies.

Close Encounter

Radio waves, visible light, gamma rays—the only difference is wavelength. But it is quite a difference. Some of the radio waves received by the Very Large Array radio telescope in New Mexico have wavelengths as large as a yardstick, whereas gamma rays have wavelengths about the size of an atomic nucleus.

The energy of a particular wavelength of electromagnetic radiation is directly proportional to its frequency. Thus, photons of light that have high frequencies and short wavelengths (such as x-rays and gamma rays) are the most energetic, and photons that have low frequencies (radio waves) are the least energetic. Ever wonder why the origin stories of comic-book superheroes often involve gamma rays? Those gamma-ray photons carry a lot of energy, and it takes a lot of energy to make a superhero, apparently.

Because we are mere mortals, however, we are highly fortunate that Earth's atmosphere absorbs most of the high-energy photons that strike it. Energetic photons tend to scramble genetic material, and the human race wouldn't last long without the protective blanket of ozone in our upper atmosphere. If a massive star were to explode somewhere near Earth in the future, the most harmful effect would be the high-energy photons that would cause a wave of mutations in the next generation.

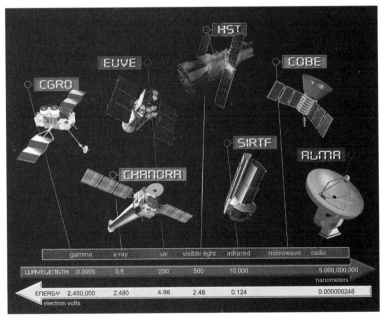

Different telescopes are sensitive to different parts of the electromagnetic spectrum.

(Image from NASA/CXC)

What Makes Color?

Visible light, we have said, is defined as light with wavelengths from 400 to 700 nm in length, about the dimensions of an average-sized bacterium.

All colors are contained within this tiny range of wavelengths. Just as wavelength (or frequency) determines whether electromagnetic radiation is visible light or x-rays or something else, so it determines what color we see. Our eyes respond differently to electromagnetic waves of different wavelengths. Red light, at the low-frequency end of the visible spectrum, has a wavelength of about 7.0×10^{-7} meters (and a frequency of 4.3×10^{14} Hz). Violet light, at the high-frequency end of the visible spectrum, has a wavelength of 4.0×10^{-7} meters (and a frequency of 7.5×10^{14} Hz). All of the other colors fall between these extremes, in the familiar order of the rainbow: orange just above red; yellow above orange; then green, blue, indigo, and violet. An easy way to remember this is to think of the name "Roy G. Biv," with every letter in the name representing the first name of a color.

So-called white light is a combination of all the colors of the visible spectrum. Light is different from paint. Dump together a lot of different colors of paint and you'll end up with a brownish gray. However, dump together all of the colors of the rainbow and you'll get "white light." When sunlight is refracted (or bent) as it passes through water droplets in the air, different colors of light are bent to different degrees. The "divided light" then reflects on the back side of the raindrop, and you see the result of this process occurring in a myriad of water droplets as a spectacular rainbow.

Astro Byte

For 25 years, cosmic gamma ray bursts (GRBs) have been one of the great mysteries of modern astronomy. GRBs have given us several clues as to what they might be. First discovered by Earth-monitoring satellites looking for secret nuclear test explosions, GRBs were seen to occur frequently and appeared to be spread evenly over the sky. Their distribution on the sky indicated that they were the result of events happening either very close, or very far (in other galaxies). The events were not, for example, seen to be concentrated in the plane of our galaxy.

Concentrated study and follow-up observations have shown that GRBs appear to arise in distant galaxies. If they are distant and very bright, then the source of the GRB must be a very energetic event. One possible explanation of GRBs is that they are the result of the merger of two neutron stars, the dense remnant cores of exhausted massive stars. This explanation has accounted for many of the known characteristics of GRBs.

Heavenly Scoop

As we said earlier in this chapter and in Chapter 3, we see celestial objects because they are producing energy, and that energy is transmitted to us in the form of electromagnetic radiation. As you will see in later chapters, different physical processes produce different wavelengths (energies) of light. Thus the portion of the spectrum from which we receive light itself is an important piece of information.

Atmospheric Ceilings and Skylights

The information—the news—we get from space is censored by the several layers of Earth's atmosphere. In effect, Earth is surrounded by a ceiling pierced by two skylights. A rather broad range of radio waves readily penetrates our atmosphere, as does a portion of infrared and most visible light, in addition to a small portion of ultraviolet. Astronomers speak of the atmosphere's *radio window* and *optical window*, which allow passage of electromagnetic radiation of these types. To the rest of the spectrum—lower-frequency radio waves, some lower-frequency infrared, and, fortunately for our survival, most of the energetic ultraviolet rays, x-rays, and gamma rays—the atmosphere is opaque, an impenetrable ceiling. An atmosphere opaque to these wavelengths, but transparent to visible light and some infrared, is a big reason why life can survive at all on Earth.

For astronomers, however, there is a downside to the selective opacity of Earth's atmosphere. Observations of ultraviolet, x-ray, and gamma-ray radiation cannot be made from the surface of Earth, but must be made by means of satellites, which are placed in orbit well above the atmosphere. It's no wonder that the advent of the space age has led to such an explosion in the amount of information that we have about the universe.

With all that energy, x-rays are hard to focus. The focusing system used by the Chandra X-ray Observatory is shown here. The energetic rays are slowly focused over a 10-m length.

(Image from NASA/CXC)

Close Encounter

Has this ever happened to you? You're on a cross-country car trip, pass through a city, and find an FM radio station you really like, only to lose it within a few miles. On the other hand, many AM stations (with programming that can range from dull to down-right frightening) seem to go on forever. The reason? FM waves are at a higher frequency and correspondingly shorter wavelength than AM. Earth's atmosphere is usually transparent to high frequency FM radio waves, but opaque to lower frequency AM. As a result, FM signals do not follow the curvature of Earth, but go off on a tangent and into outer space, whereas AM signals can be reflected downward by the atmosphere and tend to follow Earth's curve for a fair distance. If alien life forms are picking up our radio signals, they're getting mostly FM and television (which broadcasts in the upper end of the FM band), but not AM. They are, therefore, mostly denied the "pleasure" of Rush Limbaugh and Howard Stern.

The Black-Body Spectrum

As Maxwell first described in the nineteenth century, all objects emit radiation at all times because the charged atomic particles of which they are made are constantly in random motion. As these particles move, they generate electromagnetic waves. Heat an object, and its atomic particles will move more rapidly, thereby emitting more radiation. Cool an object, and the particles will slow down, emitting proportionately less and lower-energy electromagnetic radiation. If we can study the spectrum (that is, the intensity of light from a variety of wavelengths) of the electromagnetic radiation emitted by an object, we can understand a lot about the source. One of the most important quantities we can determine is its temperature. Fortunately, we don't need to stick a thermometer in a star to see how hot it is. All we have to do is look at its light carefully.

Here's how.

All objects emit radiation, but no natural object emits all of its radiation at a single frequency. Typically, the radiation is spread out over a range of frequencies. If we can determine how the *intensity* (amount or strength) of the radiation emitted by an object is distributed across the spectrum, we can learn a great deal about the object's properties, including its temperature.

Physicists often refer to a *black body*, an imaginary object that absorbs all radiation falling upon it and re-emits all the radiation that it absorbs. The way in which this re-emitted energy is distributed across the range of the spectrum is drawn as a *black-body curve*.

Now, no actual object in the physical world absorbs and radiates in this ideal fashion, but the black-body curve can be used as a model against which the peak intensity of radiation from real objects can be measured. The reason is that the peak of the black-body curve shifts toward higher frequencies (and shorter wavelengths) as an object's temperature increases. Thus, an object or region that is emitting mostly very short wavelength gamma-ray photons must be much hotter than one producing mostly longer wavelength radio waves. If we can determine the wavelengths of the peak of an object's electromagnetic radiation emissions, we can determine its temperature.

Star Words

A **black body** is an idealized (imaginary) object that absorbs all radiation that falls on it and perfectly re-emits all radiation that it absorbs. The spectrum that such an object emits is an idealized mathematical construct called a **black-body curve,** which can serve as a model for measuring the peak intensity of radiation emitted by a real object, such as a star.

Astronomers measure peak intensity with sophisticated scientific instruments, but we all do this intuitively almost every day. You have an electric kitchen range, let's say. The knob for one of the heating elements is turned to *off*. The heating element is black in color. This tells you that it might be safe to touch it. But if you were to turn on the element, and hold your hand above it, you would feel heat rising, and would know that it was starting to get hot. If you had infrared vision, you would see the element "glowing" in the infrared. As the element grows hotter, it will eventually glow visibly red, and, even without special infrared vision, you would know that it was absolutely a bad idea to touch it (regardless of where the control knob happened to be pointing).

At room temperature, the metal of the heating element is black, but as it heats up, it changes color: from dull red to bright red. If you had a very high-voltage electric range and a sufficiently durable heating element, you could crank up the temperature so that it became even hotter. It would emit most of its electromagnetic radiation at progressively higher frequencies.

Now, an object that omits *most* of its radiation at optical frequencies would be *very* hot. And a kitchen range will never (we hope) reach temperatures of 6,000 K, like the Sun. The red color you see from the range is in the "tail" of its black-body spectrum. Even when hot, it is still emitting most of its radiation in the infrared part of the spectrum.

Watch Your Head! Here Comes an Equation

Physicists have boiled this black-body curve business down to a couple of laws. Wien's Law states that the wavelength of peak emission is proportional (the symbol α means "is proportional to") to 1 over absolute temperature as expressed in Kelvins:

$$\text{Wavelength of peak emission } \alpha \; \frac{1}{\text{Temperature}}$$

This law puts in equation form what we have been saying in words. As something gets hotter, its peak wavelength gets smaller. The two are inversely proportional. Don't worry, we'll discuss absolute temperature and *Kelvins* in Chapter 12. For now, just know that the hotter an object, the bluer its radiation, and the cooler an object, the redder its radiation.

It is also true that the total energy an object radiates is proportional to its temperature—in fact, it is proportional to the fourth power of the object's temperature:

$$\text{Total energy radiated } \alpha \; \text{Temperature}^4$$

In other words, if you double the temperature of an object, say from 1,000 K to 2,000 K, it will radiate not twice the energy, but *16* times the energy (2^4, or $2 \times 2 \times 2 \times 2 = 16$). This relationship between energy radiated and temperature is called the Stefan-Boltzmann Law.

> **Star Words**
>
> The **Kelvin** (K) temperature scale is tied to the Celsius (C) temperature scale, and is useful because there are no negative Kelvin temperatures. 0° C is the temperature at which water at atmospheric pressure freezes. 100° C is the temperature at which water boils. Absolute zero (0° K) is the coldest temperature that matter can attain. At this temperature, the atoms in matter would stop jiggling around all together. 0° K corresponds to approximately −273° C, and under laboratory conditions, temperatures in the milliKelvin (thousandths of a Kelvin) range can be attained. For a sense of scale, stars like the Sun have surface temperatures of about 6,000° K. A warm day on Earth is about 300° K.

Read Any Good Spectral Lines Lately?

Using the spectrum and armed with the proper instrumentation, astronomers can accurately read the temperature of even very distant objects in space. And even without sophisticated equipment, you can startle your friends by letting them know that Betelgeuse (a *reddish* star) must have a lower surface temperature than the *yellow* Sun.

Astronomers also use the spectrum to learn even more about distant sources. A *spectroscope* passes incoming light through a narrow slit and prism, splitting the light into its component colors. Certain processes in atoms and molecules give rise to emission at very particular wavelengths. Using such a device, astronomers can view these individual spectral lines and glean even more information about conditions at the source of the light.

Although ordinary white light simply breaks down into a continuous spectrum—the entire rainbow of hues, from red to violet, shading into one another—light emitted by certain substances produces an emission spectrum with discrete *emission lines*, which are, in effect, the fingerprint of the substance.

Hydrogen, for example, has four clearly observable spectral lines in the visible part of the spectrum (red, blue-green, violet, and deep violet). The color from these four lines (added together as light) is pinkish. These four spectral lines result from the electron that is bound to the proton in a hydrogen atom jumping between particular energy levels. There are many other spectral lines being emitted; it just so happens that only four of them are in the visible part of the spectrum. Hot hydrogen gas is the source of the pinkish emission from regions around young stars like the Orion nebula (see Chapter 14).

> **Star Words**
>
> A **spectroscope** is an instrument that passes incoming light through a slit or prism, splitting it into its component colors. All substances produce characteristic **spectral lines**, which act as the "fingerprint" of the substance, enabling identification of it.

> **Star Words**
>
> **Emission lines** are narrow regions of the spectrum where a particular substance is observed to emit its energy. Atomic emission lines result from basic processes occurring on the smallest scales in an atom, electrons bound to a nucleus moving between different energy states.

In our hydrogen atom example, a negative electron is bound to a positive proton. The electron, while bound to the proton, can only exist in certain specific states or *energy levels*. Think of these energy levels as rungs on a ladder. The electron is either on the first rung, the second rung, or the third rung, and so on … it can't be in between. When the electron moves from a higher energy level to lower one, it gives off energy in the form of a photon. Because the levels that the electron can inhabit are limited, only photons of a few specific frequencies are given off. These particular photons are apparent as bright regions in the spectrum of hydrogen: the element's spectral emission lines.

This simulated x-ray spectrum of a cluster of galaxies shows the combination of a black-body spectrum—the big curve—with emission lines superimposed on it. Spectra contain a lot of information beyond that in a normal image.

(Image from NASA/CXC)

Astro Byte

Spectral lines have a certain rest frequency; that is, there is a certain frequency at which we expect to see the line. If the line is somewhere else, or if a series of lines are all systematically moved, then the source of the lines might be in motion relative to us. If the source is coming toward us, the lines will occur at a higher frequency (known as a blue shift), and if the source is moving away from us, the lines will occur at a lower frequency (known as a red shift). The Doppler shift is most familiar in sound waves. We have all heard the shift in tone (frequency) that occurs when a car drives past us, especially if the driver leans on the horn as he passes. Doppler shifts are a property of all waves, whether light or sound. This effect has been used to discover planets around nearby stars (see Chapter 11 for more on this).

Depending on how you view an astronomical source, you will see different types of spectra. A black-body source that is viewed directly will produce a continuous spectrum. But if the photons from the source pass through a foreground cloud of material, certain energies will be absorbed by the cloud (depending on its composition), and you will see a black-body spectrum with certain portions of the spectrum missing or dark. This is called an absorption spectrum. If a cloud of material absorbs energy and then re-emits it in a different direction, you see the result as emission lines, or bright regions in the spectrum. The clouds of hot gas around young stars produce such emission lines.

The light that reaches us from stars carries a lot of information. The color of the object can tell you its temperature, the wavelength of the light reaching you can tell you about the energies involved, and the presence (or absence) of certain wavelengths in a black-body spectrum can tell you what elements are present in a given source.

And all of this information comes to us as we observe the Universe from the comfort of our home planet. Who knew we could learn so much without actually going anywhere?

The Least You Need to Know

- Visible light is a rather small subset of the electromagnetic spectrum.

- The difference between visible light and other electromagnetic waves, say, radio waves, is just a matter of wavelength or frequency.

- Unlike sound waves, light waves can travel through a vacuum—empty space—because these waves are disturbances in the electromagnetic field and require no medium (substance) for transmission.

- The peak wavelength in an object's spectrum tells us its temperature.

- One way spectral lines arise is by the specific energies given off when electrons jump between energy levels in an atom, or when molecules spin at different rates.

- Astronomers use spectroscopes to read the spectral "fingerprint" produced by the light received from distant objects and thereby determine the chemical makeup of an object.

Chapter 5

Seeing in the Dark

In This Chapter

- ◆ How radio telescopes work
- ◆ Interferometers: what they are and what they do
- ◆ The purpose of radio astronomy
- ◆ Amateur radio astronomy
- ◆ Monitoring solar flares and Jupiter's radio noise
- ◆ Other nonvisible astronomy techniques

"What's an astronomer, Daddy?"

Spending much time around a little boy or girl can be pretty exhausting. All those questions! At least this one has a quick answer: "An astronomer is a person who looks at the sky through a telescope."

"But, Daddy, the visible spectrum is squeezed between 400 and 700 nm. What about the rest of the electromagnetic spectrum?"

Smart kid.

Until well into the twentieth century, astronomers had no way to "see" most of the nonvisible electromagnetic radiation that reached Earth from the universe. Then along came radio astronomy, which got its start, quite by accident, in 1931–1932 and was cranking into high gear by the end of the 1950s. Over the past 40 years or so, much of our current knowledge of the universe has come about through radio observations.

Radio astronomy is simply the study of the universe at radio wavelengths. Astronomers used to categorize themselves by the wavelength of the observations that they made: radio astronomer versus optical astronomer. Increasingly, though, astronomers define their work more by what they study (pulsars, star formation, galactic evolution) than by what wavelength they use. The reason for this change is that, in recent years, new instruments have opened the electromagnetic spectrum to an unprecedented degree. Astronomers now have the ability to ask questions that can be answered with observations at many different wavelengths.

This chapter explains what professional radio astronomers do and how amateurs can make their own observations in this low frequency part of the spectrum.

Dark Doesn't Mean You Can't See

On a clear night far from urban light pollution, the sky is indeed dazzling. Just remember that the electromagnetic information your eyes are taking in, wondrous as it is, comes from a very thin slice of the entire spectrum. As we mentioned in the last chapter, Earth's atmosphere screens out much of the electromagnetic radiation that comes from space. It allows only visible light and a bit of infrared and ultraviolet radiation to pass through a so-called optical window, but it allows a broad portion of the radio spectrum to pass through a radio window.

Two windows.

If your house had two windows, would you look through only one?

A Telephone Man Tunes In

The first true radio astronomer wasn't trained as an astronomer at all. Even to this day, many astronomers who work in the radio regime were trained as physicists and electrical engineers, and only later learned to apply their knowledge to astronomy. Karl Jansky, the son of a Czech immigrant who settled in Oklahoma (where Karl was born in 1905), took a degree in physics at the University of Wisconsin. After graduating, Karl went to work as a telephone engineer with Bell Labs. The phone company was looking for ways to make telephone communications possible with shortwave radio, but the transmissions were bedeviled by all sorts of interference.

Now most people hadn't given much thought to radio static. After all, static was something to be avoided if at all possible—meaningless noise that only interfered with communications. Jansky was given the assignment of studying sources of static at a wavelength of 14.6 m in an effort to track down the precise sources of radio interference and eliminate them.

Karl Jansky, shown here during the early 1930s with his merry-go-round radio antenna, which first detected radio waves from the center of our Milky Way galaxy.

(Image courtesy of AUI/NRAO)

On a farm in Holmdel, New Jersey, not far from Bell Labs, Jansky set up a very un-gainly looking device, which he called a merry-go-round. It was a large directional antenna, which looked rather like the biplane wing of the Wright brothers' first air-plane. It was mounted on some discarded Model T Ford wheels and could be rotated 360 degrees by means of a motor. Using this contraption, Jansky was soon able to identify all the known sources of radio interference except one.

Jansky tracked the stubborn and mysterious interference. When amplified and sent to a speaker, the interference sounded like a faint hiss. The source *seemed* to be in the sky, because Jansky could track it rising and setting with the stars.

But it wasn't coming from just anywhere in the sky. By the spring of 1932, Jansky traced the primary source of radio noise to the direction of the constellation Sagittarius, which astronomers Harlow Shapley and Jan H. Oort had identified (from the distribution of globular clusters in the Galaxy) as the direction of the center of our home Milky Way galaxy. Using his merry-go-round antenna, Jansky had actually "listened in on" the center of the much bigger merry-go-round that is our Galaxy. There were other sources of radio noise in the sky as well, but Jansky noted that the Sun itself was not an impressive source of radio noise. This observation was a bit surprising, because the Sun is so close to us, but, based on the observation, he concluded that whatever the source of radio noise, it probably wasn't stars.

Jansky published his "discovery" late in 1932, and the detection of radio signals from space appeared in national newspapers by the following year. Strangely enough, Jansky himself didn't pursue the science he had accidentally created. As for most professional astronomers, they continued to look through only one of their two windows, the portion of the spectrum available to optical telescopes.

It took another nonastronomer, Grote Reber, to appreciate the possibilities of what Jansky had discovered. In today's image-conscious world, we might call Reber a nerd. But, as the example of Bill Gates has shown us, some nerds go on to change the world. Born in Wheaton, Illinois, in 1911, he grew up tinkering with radio transmitters, building one powerful enough to communicate with other ham radio operators all over the world. He became an electrical engineer, but never lost his interest in amateur radio, and when he read about Jansky's discovery, he tried, without success at first, to adapt his own short wavelength receiver to pick up interstellar radio waves with wavelengths of 10 cm.

> **Astro Byte**
>
> Although Jansky didn't follow up on his discovery, he is considered the father of radio astronomy. Radio astronomers have honored him by naming the basic unit of radio brightness the Jansky (Jy).

> **Close Encounter**
>
> Until after World War II, Grote Reber's 30-foot dish was the only radio telescope in the world, and by 1942 he had completed the world's first preliminary radio maps of the sky. The end of WWII started a period of intense research in radio astronomy all over the world. A large number of former radar engineers (particularly in England and Australia) turned their talents to the pursuit of more distant sources using radio telescopes. After the war, Reber moved his radio telescope to Sterling, Virginia, and was given a government job in Washington, D.C., as chief of the Experimental Microwave Research Section. In 1951, he built a new radio telescope on an extinct Hawaiian volcano and mapped out low-frequency (long-wave) celestial signals (previously he had concentrated on shortwave signals). In pursuit of a bigger radio window, he moved in 1954 to Tasmania, Australia, a place where Earth's atmosphere is occasionally transparent to electromagnetic radiation more than 30 meters in wavelength. In 1957 he returned for a time to the United States to work at the newly created National Radio Astronomy Observatory in Green Bank, West Virginia, but went back to Tasmania in 1961 to continue mapping long-wave radio sources. Reber died in 2002.

He tinkered with the electronics (trying longer wavelengths), and, in 1937, built a paraboloidal antenna 30 feet in diameter. With this, Reber not only confirmed Jansky's discovery of radio waves from the direction of Sagittarius, but he found other sources in the direction of the constellations Cygnus, Cassiopeia, and elsewhere.

Reber confirmed Jansky's conclusion that the radio signals didn't coincide with the positions of visible stars. Directing his dish toward such bright stars as Sirius, Vega, and Rigel, he detected nothing. But looking toward a starless area in Cassiopeia, he picked up strong radio waves. He had unknowingly detected a supernova remnant known as Cassiopeia A.

Cassiopeia A is the brightest radio source in the constellation Cassiopeia. A supernova remnant, it consists of the gaseous debris of a supernova explosion that occurred in the late seventeenth century.

(Image courtesy of AUI/NRAO)

Anatomy of a Radio Telescope

The basic anatomy of a radio telescope hasn't changed all that much from Reber's dish—although the instruments have become much larger and the electronics far more sophisticated. In principle, a radio telescope works like an optical telescope. It is just a "bucket" that collects radio frequency waves rather than visible light waves, and focuses them on a detector or receiver. A large metal dish—like a giant TV satellite dish—is supported on a moveable mount. A detector, called a receiver horn, is mounted on legs above the dish (prime focus) or below the surface of the dish (Cassegrain focus). The telescope is pointed toward the radio source, and its huge dish collects the radio waves and focuses them on the receiver, which amplifies the signal and sends it to a computer.

Because the radio spectrum is so broad, astronomers have to decide which portion of the radio spectrum to observe. Different receivers are used for observations at different frequencies. Receivers are either swapped in and out, or (more typically) the radio signal is directed to the correct receiver by moving a secondary reflecting surface (like the secondary mirror in an optical telescope).

Astronomer's Notebook

Why do radio telescopes have to be so big? A 30-foot or 100-meter dish might seem excessive. The resolution is actually determined by the ratio of the wavelength being observed to the diameter of the telescope. So optical telescopes (which detect short-wavelength optical photons) can be much smaller than radio telescopes, which are trying to detect long-wavelength radio waves, and have the same resolution.

Bigger Is Better: The Green Bank Telescope

In the case of radio telescopes, size really does matter. The resolution of a telescope depends not only on its diameter, but the wavelength of the detected radiation (the ratio of wavelength to telescope diameter determines the resolution). Radio waves are big (on the order of centimeters or meters), and the telescopes that detect them are correspondingly huge. Also, the radio signals that these instruments detect are very faint, and just as bigger optical telescopes with bigger mirrors collect more light than smaller ones, bigger radio telescopes collect more radio waves and image fainter radio signals than smaller ones.

Collecting radio signals is just part of the task, however. You might recall from Chapter 3, that for practical purposes, very good optical telescopes located on Earth's surface can resolve celestial objects to 1" (1 arcsecond—$\frac{1}{60}$ of 1 arcminute, which, in turn, is $\frac{1}{60}$ of 1 degree). The best angular resolution that a very large single-dish radio telescope can achieve is some 10 times coarser than this, about 10", and this, coarse as it is, is possible only with the very largest single dish radio telescopes in the world. The National Radio Astronomy Observatory has recently completed the world's largest fully steerable radio telescope. The 100-m dish of the Green Bank Telescope at Green Bank in Pocahontas County, Virginia, has a best resolution of 14".

The world's largest nonsteerable single-dish radio telescope was built in 1963 at Arecibo, Puerto Rico, and has a dish 300 meters (984 feet) in diameter sunk into a natural valley. Although its great size makes this the most sensitive radio telescope, the primary surface is nonsteerable—totally immobile—and, therefore, is limited to observing objects that happen to pass roughly overhead (within 20 degrees of zenith) as Earth rotates.

The Arecibo Radio Telescope, Arecibo, Puerto Rico.

(Image from National Astronomy and Ionosphere Center)

Interference Can Be a Good Thing

There is a way to overcome the low angular resolution due to the size of radio waves: link together a lot of smaller telescopes so they act like one giant telescope. A *radio interferometer* is a combination of two or more radio telescopes linked together electronically to form a kind of virtual dish, an array of antennas that acts like one gigantic antenna. It is as if we had small pieces of a very large optical telescope (imagine a giant mirror with a lot of its surface area punched out), so that although an interferometer has the resolution of a very large telescope, it does not have the surface area or sensitivity to faint sources of a truly gigantic telescope.

Star Words _____

A **radio interferometer** is a combination of radio telescopes linked together electronically to create the equivalent (in terms of resolution) of a giant radio telescope. The use of an interferometer greatly increases resolving power of radio frequencies. Earth-based radio interferometers have much better resolution than any conventional Earth-based optical telescope. Of course, optical interferometers (which are being tested) will exceed radio interferometers in resolution when they go into operation because of the higher frequency of optical radiation.

The National Radio Astronomy Observatory (NRAO) maintains and operates the Very Large Array (VLA) interferometer on a vast plain near Socorro, New Mexico, consisting of 27 large dishes arrayed on railroad tracks that are laid out in a Y-shaped pattern. Each arm is 12.4 miles (20 km) long, and the largest distance between two of the antennas is 21.7 miles (35 km). As a result, the VLA has the resolving power of a radio telescope 21.7 miles across.

For radio astronomers who want something even larger than "very large," there is Very Long Baseline Interferometry (VLBI), which can link radio telescopes in different parts of the world to achieve incredible angular resolutions of a thousandth of an arcsecond (.001" or better. From its offices in Socorro, New Mexico, the NRAO also operates the VLBA (Very Long Baseline Array), which consists of 10 radio dishes scattered over the United States, from Mauna Kea, Hawaii, to St. Croix, U.S. Virgin Islands.

Still not big enough? In 1996, Japanese astronomers launched into Earth orbit a radio telescope to be used in conjunction with the ground-based telescopes in order to achieve the resolution of a telescope larger than Earth itself. For more information, see www.nrao.edu.

What Radio Astronomers "See"

Insomnia is a valuable affliction for optical astronomers, who need to make good use of the hours of darkness when the Sun is on the other side of Earth. But as Karl Jansky discovered so many years ago, the Sun is not a particularly bright radio source and therefore won't interfere with observations in the radio spectrum. In consequence, radio astronomers (and radio telescopes) can work night and day. The VLA, for example, gathers data (or runs tests) 24 hours a day, 363 days a year. Not only is darkness not required, but you can make radio observations through a cloud-filled sky. The senior author of this book even observed a distant star-forming region in the midst of a storm during which lightning struck near the VLA and disabled it for a few minutes.

As the Dutch astronomer Jan Oort realized after reading Reber's work in the 1940s, radio waves opened new vistas into the Milky Way and beyond. Radio astronomers can observe objects whose visible light doesn't reach Earth because of obscuration by interstellar dust or simply because they emit little or no visible light. The fantastic objects known as quasars, pulsars, and the regions around black holes—all of which you will encounter later in this book—are often faint or invisible optically, but do emit radio waves.

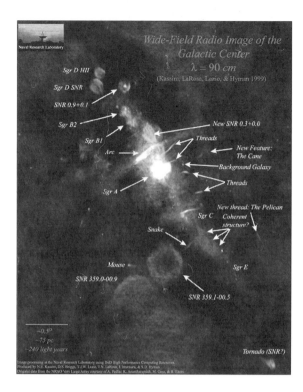

Image of the Galactic center as seen with the Very Large Array (VLA) at 90 cm.

(Image from N. Kassim et al., NRL)

The spiral form of our own Galaxy was first mapped using the 21-cm radio spectral line from neutral (cold) hydrogen atoms, and the discovery of complex molecules between the stars was made at radio frequencies. The very center of our own Milky Way galaxy is hidden from optical probing, so that most of what we know of our galactic center has come from infrared and radio observation. Because radio interferometers detect an interference pattern, radio data has to be processed in ways different from optical data. But the end result is either a radio image, showing the brightness of the source in the sky, or a radio spectrum, showing a spectral line or lines. In Chapter 14, we'll describe how radio spectral lines arise.

Do-It-Yourself Radio Astronomy: No-Cost and Low-Cost Approaches

Building a huge radio telescope like Arecibo or the Green Bank Telescope (GBT) takes a great deal of money, and so does operating one. Even if you had the cash, your neighbors (not to mention the local zoning board) might frown on your building even a modest 30-foot-diameter dish antenna in your backyard. But remember that radio astronomy originated with nonastronomers, and there is still plenty of room in the field for amateurs, including amateurs of modest means. A small but committed group of enthusiasts have formed the Society of Amateur Radio Astronomers (SARA).

Astronomer's Notebook

You can contact the Society of Amateur Radio Astronomers (SARA) at its website: www.qsl.net/SARA. The website of Radio-Sky Publishing also offers valuable information on amateur radio astronomy: www.radiosky.com.

Astronomer's Notebook

You don't have to wait for clear skies or night to fall to detect meteor events, but meteor showers and even out-and-out meteor storms do occur at certain times of the year, when Earth passes through the debris that litters the "orbital paths" of known comets. Two of the most intense are the Perseid shower in August and the Geminid shower in December.

If you own an FM radio or a television set, you can do some radio astronomy at no cost whatsoever (we thank Tom Crowley of Atlanta Astronomy Club for many of the following ideas).

Have you ever witnessed a meteor shower? The streaks of light in the night sky can be quite spectacular. Meteors are the bright trails of ionized atmosphere behind tiny bits of cometary debris that enter Earth's atmosphere. Most meteorites are no larger than a pea. What if we told you that there was another way to watch a meteor shower—by using your FM radio as a radio telescope?

Meteor counts by radio are about 10 times more accurate than visual observation—and, as with any radio observations, you can observe them day or night, rain or shine. Recall from Chapter 4 that Earth's atmosphere is transparent to some forms of electromagnetic radiation and opaque to others. The upper atmosphere normally reflects low-frequency AM radio signals. In contrast, the atmosphere is transparent to higher-frequency FM radio waves, which, as a consequence, have a shorter range. They usually penetrate and are not reflected by the atmosphere.

But something happens when a meteorite enters the atmosphere. Each piece of debris that tears into the atmosphere (at up to 40 miles per second) heats up the air around it and creates a tiny ionized (electrically charged) vapor trail in the upper atmosphere. These columns of charged particles can reflect even higher-frequency FM radio waves. This temporary condition means that previously out-of-range FM broadcasts can (for a moment) be heard. During periods of known meteoric activity, stay at the low end of the FM dial and try to find FM radio stations that are from 400 to 1,300 miles away. You might call a distant friend to get the broadcast frequency of a few stations. When a distant station fades in for a second or two, you are indirectly observing a meteor. The ionized trail behind it has momentarily reflected a distant radio signal into your receiver.

It helps if you can hook up your radio to an outdoor antenna, but if the meteor shower is fairly intense, you should detect many events even without such an antenna. You can also try tuning your TV set to the lowest unused VHF channel. Again, when a distant station, normally out of range, fades in and the signal becomes strong for a second or two, you know a meteor has entered the atmosphere. (Note that this works only with a television receiving signals from an outdoor antenna—not via cable or satellite!)

Just by tuning in your radio, you can do some meaningful radio astronomy. You can make it more interesting by recording the events on tape, or keeping a written record of the number of events you detect per hour.

But amateur radio astronomy need not be limited to listening for distant FM radio or TV stations. If you are an amateur radio operator—a ham—you already have much of the equipment required for more serious radio astronomy. If you aren't into amateur radio, you can get started for a highly variable but modest cost. The first step to take is to log onto SARA's World Wide Web site (www.qsl.net/SARA) for overview information.

Essentially, amateurs can use either nonimaging or imaging radio astronomy techniques. Nonimaging techniques (which monitor radio emissions without pinpointing locations) require a simple shortwave receiver, usually modified to receive a narrow band of frequencies, and a simple antenna system. With such equipment, you can track radio emissions from Jupiter, *solar flares*, and meteor events.

Imaging techniques (which provide more detailed information on the location and nature of the signal) require a more serious commitment of resources, including a much larger dish-type antenna, more sophisticated receiving equipment, reasonably elaborate recording equipment, and (probably) a rural location removed from most sources of terrestrial radio interference. We'll stick to the more approachable nonimaging techniques, which are more appropriate for beginners.

Star Words

Solar flares are explosive events that occur in or near an active region on the Sun.

Solar Flares and Meteor Events

Solar flares, which we'll discuss in Chapter 12, are explosive events that occur in or near an active region on the Sun's surface. Flares can be detected with a very low-frequency (VLF) receiver operating in the 20- to 100-KHz (kilohertz, or thousand hertz) band. (This is below the region of the spectrum where AM radio stations broadcast.) Such a receiver can be homemade, using plans supplied by such organizations as SARA. Solar flares can also be monitored in regular shortwave radio bands with a standard shortwave receiver.

Why do solar flares create radio signals? They generate x-rays that strike a part of Earth's atmosphere called the ionosphere, greatly enhancing the electron count in this atmospheric region. These electrons generate the noise picked up by the radio.

Solar flares give off energy across the electromagnetic spectrum. Pictured is a bright solar flare extending from the surface of the Sun, as imaged at 195 Angstroms (Å) in 1998. When solar flares occur, material is launched outward at millions of km/h.

(Image courtesy of SOHO)

ET Phone Home

If you've seen such sci-fi movies as *The Arrival, Independence Day,* or *Contact,* or have read Mary Doria Russel's novel *The Sparrow,* you already know about an organization called SETI (Search for Extra-Terrestrial Intelligence). It is an international group of scientists and others who, for the most part, use radio telescopes to monitor the heavens in search of radio signals generated not by natural phenomena, but broadcast artificially by intelligent beings from other worlds. So far, no clearly artificial extraterrestrial radio signals have been confirmed, but SETI personnel keep searching. The SETI project got a large boost in August 2000 when Paul Allen, one of the co-founders of Microsoft, and Nathan Myhrvold, Microsoft's former chief technology officer, committed $12.5 million to the project for the design and construction of the Allen Telescope Array, an instrument dedicated to the exclusive use of SETI.

We'll take a closer look at SETI and other efforts to make contact in Chapter 22, but if you're interested in the search for extraterrestrial intelligence, you can do more than just read about it.

A SETI project called SETI@home offers as a free download to any computer user a special kind of screensaver program, which (as of Fall 2003) had been installed on over 4,648,655 computers in more than 224 countries. This screensaver does more than put up a pretty picture when your computer is idle. It *uses* that idle time to go to work analyzing data from four million different combinations of frequency bandwidth and drift rate recorded at the Arecibo radio telescope. The data is divided and uploaded to millions of SETI@home users, whose computers crunch the numbers. By means of such "distributed computing," the SETI Institute can analyze data more quickly, thereby increasing the chances of ferreting out a radio signal from an intelligent source. If you want to download the screensaver and join the quest for ET—or if you just want to learn more—point your browser to http://setiathome.ssl.berkeley.edu/.

The Rest of the Spectrum

Optical astronomy with the naked eye is at least 5,000 years old and probably much older. Optical telescope astronomy is about 400 years old. Radio astronomy is a youthful discipline at about 70 years, if we date its birth from Jansky's work in the early 1930s. But it has been only since the 1970s that other parts of the electromagnetic spectrum have been regularly explored for the astronomical information they might yield. Each new window thrown open on the cosmos has brought in a fresh breeze and enriched our understanding of the universe.

New Infrared and Ultraviolet Observations

Telescopes need to be specially equipped to detect infrared radiation—the portion of the spectrum just below the red end of visible light. Infrared observatories have applications in almost all areas of astronomy, from the study of star formation, cool stars, and the center of the Milky Way, to active galaxies, and the large-scale structure of the universe.

IRAS (the Infrared Astronomy Satellite) was launched in 1983 and sent images back to Earth for many years. Like all infrared detectors, though, the ones on IRAS had to be cooled to low temperatures so that their own heat did not overwhelm the weak signals that they were trying to detect. Although the satellite is still in orbit, it has long since run out of coolant, and can no longer make images. The infrared capability of the *Hubble Space Telescope* provided by NICMOS (Near-Infrared Camera and Multi-Object Spectograph) yielded spectacular results while in operation. Launched on August 25, 2003, the Space Infrared Telescope Facility (SIRTF) is just starting its 2.5-year mission, observing the universe at wavelengths between 3 and 180 microns (a micron is a millionth of a meter).

And NASA's *James Webb Space Telescope* (formerly the *Next-Generation Space Telescope*—NGST), scheduled for launch in 2011, will be optimized to operate at infrared wavelengths, and will be cooled passively by a large solar shield.

Ultraviolet radiation, which begins in the spectrum at frequencies higher than those of visible light, is also being studied with new telescopes. Because our atmosphere blocks all but a small amount of ultraviolet radiation, ultraviolet studies must be made by high-altitude balloons, rockets, or satellites like the *Hopkins Ultraviolet Telescope* (HUT) flown in 1990 and 1995 in the payload bay of the space shuttle *Endeavour*.

The *Hubble Space Telescope* also has the capability to detect ultraviolet (UV) photons. Ultraviolet observations provide our best views of stars, particularly stars with surface temperatures higher than the Sun's.

Chandrasekhar and the X-Ray Revolution

Astronomers can now study electromagnetic radiation at the highest end of the spectrum. Because x-rays and gamma rays cannot penetrate our atmosphere, all of this work must be done by satellite. X-rays are detected from very high-energy sources, such as the remnants of exploded stars (supernova remnants) and jets of material streaming from the centers of galaxies. Work began in earnest in 1978 with the launch of the *High-Energy Astronomy Observatory* (later dubbed the Einstein Observatory), an x-ray telescope. Germany launched the next x-ray telescope, the *Röentgen Satellite* (ROSAT), in 1990. Finally, in 1999, the *Chandra X-ray Observatory* (named for astronomer Subrahmanyah Chandrasekhar), the world's premier x-ray instrument, was launched. It has produced unparalleled high-resolution images of the x-ray universe. The Chandra image of the Crab Nebula, home to a known pulsar, showed never before seen details of the environment of an exploded star. For recent images, go to www.chandra.harvard.edu, which is updated continuously.

In 1991, the *Compton Gamma Ray Observatory* (CGRO) was launched by the space shuttle and was in operation until June 2003. It revealed unique views of the cosmos, especially in regions where the energies involved are very high: near black holes, at the centers of active galaxies, and near neutron stars. We'll have more to say about many of these phenomena in Part 3, but you can check out a mission summary of the CGRO at http://cossc.gsfc.nasa.gov/.

Capturing the Full Spectrum

After a satellite is launched, any astronomer can make a proposal to obtain observing time. There are always many more proposals than time available, but this time-sharing practice nevertheless makes some of the world's most advanced technologies available to astronomers at institutions worldwide. Today's astronomers have the unique ability to turn to almost any region of the electromagnetic spectrum for answers to their questions. They can now play on the full keyboard, instead of plunking away on a single key.

The Least You Need to Know

- ◆ Electrical engineer Karl Jansky invented radio astronomy in the 1930s, when he was tracing sources of shortwave radio interference for the telephone company.

- ◆ A radio telescope typically consists of a large parabolic dish that collects and focuses very weak incoming radio signals on a receiver. The signal is then amplified and processed using electronics and computers.

- ◆ Radio telescopes can detect objects that are blocked by interstellar material or that emit little or no visible radiation.

- ◆ Although professional radio astronomy typically involves elaborate equipment and massive facilities, amateur radio astronomers can use equipment as modest as an ordinary FM radio.

- ◆ In recent years, astronomers have launched instruments into orbit that can detect all segments of the electromagnetic spectrum, from infrared, through visible, and on to ultraviolet, x-rays, and gamma rays. The highest frequency radiation (x-rays and gamma rays) comes from some of the most energetic and exotic objects in the universe.

Part 2

Worlds Without End

This part is an extended tour of the solar system, beginning with our nearest neighbor, the Moon. After some lunar exploration, we venture to an overview of the solar system (Chapter 7). We even find time to linger among the comets and the asteroids.

The next three remaining chapters of this part are devoted to the planets and the moons that cruise our "neighborhood." First, we investigate the planets most like our own, the terrestrials Mercury, Venus, and Mars. Then we go on to the distant jovian worlds—the outer giants Uranus, Neptune, Jupiter, and Saturn. In Chapter 10, we take a closer look at the moons and rings of the jovians—and at the last planet discovered in our solar system, Pluto, which, as we'll see, seems in many ways more like a moon than a planet. The final chapter in this part takes us to the newly discovered realm of extrasolar planets, worlds beyond our solar system.

Chapter 6

The Moon: Our Closest Neighbor

In This Chapter

♦ Galileo observes the Moon

♦ How to observe the Moon

♦ Peeking at the far side of the Moon

♦ How the Moon affects our tides

♦ Lunar geology

♦ Who needs the Moon?

♦ Some theories on how the Moon was formed

It has been more than three decades since Neil Armstrong stepped from the *Apollo 11 Lunar Module* onto the surface of the Moon. The Moon is still the only celestial body other than Earth where humans have stood. But why did we go there?

From the pictures we've all seen, the lunar landscape is one of rock, dust, and desolation—"magnificent desolation," *Lunar Module* pilot Buzz Aldrin called it. And although the astronauts skipped across its surface, they were clearly happy to return to mother Earth. Why in the world did our nation expend such effort, treasure, and risk to send astronauts to the lunar surface? What have we learned from the information that we brought back? Why does the Moon so fascinate us? We'll explore these issues and more in this chapter.

What If We Had No Moon?

It seems like a reasonable question to ask. What if we had no moon? Would it matter? What has the Moon done for us lately?

It turns out that the presence of a large moon such as we have is unusual for a terrestrial planet. Mercury and Venus have no moons at all, and Mars has two tiny ones, Phobos and Deimos. To have a moon roughly ⅓ the size of the planet is unique in the inner solar system. Our moon, for example, is as large as some of the moons of the giant gas planets in the outer solar system. If there were no moon, we would have no ocean tides, and the rotation rate of Earth would not have slowed to its current 24 hours. It is thought that early in its life Earth rotated once every six hours. The Moon also appears to stabilize the rotational axis of Earth. By periodically blocking the light from the Sun's photosphere, the Moon gives us a view of the outer layers of the Sun's atmosphere, and it also gave early astronomers clues to the relative locations of objects in the solar system.

Close Encounter

Did you ever wonder why the Sun and the Moon are the same size in the sky? In actual physical size, the Sun dwarfs the Moon, but the Sun is so much farther away, that the two appear the same size. Try this: Hold a dime and a quarter in front of your face. Now move them back and forth until they appear to be the same size. Which do you have to hold farther away for this to be true? The bigger one, of course! We happen to be on Earth at a time when the Moon exactly blocks the light from the Sun's photosphere during a solar eclipse. As the Moon slowly drifts away from Earth, it will get smaller and smaller in the sky, and people will be entitled eventually to grumble, "Solar eclipses just aren't what they used to be."

Lunar Looking

Although the world greeted Jules Verne's 1865 book *De la Terre à la Lune* (translated in 1873 as *From the Earth to the Moon*) with acclaim and wonder, it was hardly the first fictional speculation about a voyage to our nearest cosmic neighbor. The Greek satirist Lucian had written about such a flight as early as the second century C.E. and the Moon, our constant companion, has always been an object of intense fascination. Its reflected silvery glow bathes Earth with romance and mystery. Its changing face, as it travels through its monthly cycle, has always commanded our attention, as have its peculiarly human qualities: Unlike the stars, it has been known—especially since Galileo first observed it with a telescope in 1609—to be pocked, mottled, and imperfect. Almost all cultures at all times have seen some sort of face or figure in the features of the Moon. But only recently have we realized just how important the Moon has been in the evolution of our planet, and perhaps even life on our planet.

The Sun is so intensely brilliant that to gaze at it is to go blind. But the Moon shines with harmless reflected light that invites us to gaze and gaze—to become lunatics.

What Galileo Saw

What Galileo saw when he trained a telescope on the Moon's surface conflicted with existing theories that the surface was glassy smooth; it was instead rough and mountainous. He closely studied the *terminator* (the boundary separating day and night), and noted the shining tops of mountains. Using simple geometry, he calculated the height of some of the mountains based on the angle of the Sun and the estimated length of shadows cast. Galileo over-estimated the height of the lunar mountains he observed; but he did conclude rightly that their heights were comparable to earthly peaks.

Noticing mountains and craters on the Moon was important, because it helped Galileo conclude that the Moon was fundamentally not all that different from Earth. It had mountains, valleys, and it even had what were called seas—in Latin, *maria*, although there is no indication that Galileo or anyone else maintained after telescopic observations that the maria were water-filled oceans. Contending that the Moon resembled Earth was not a small thing in 1609, because such a statement

Star Words

The **terminator** is the boundary separating light from dark, the daytime from nighttime hemispheres of the Moon (or other planetary or lunar body). The terminator marches across the lunar surface at a very slow rate, crossing the entire surface of the Moon once every 27.3 Earth days. That's a long day.

Star Words

Maria (pronounced "MAH-ree-uh") is the plural of *mare* (pronounced "mar-ay"), Latin for "sea." Maria are dark gray plains on the lunar surface, which to earlier observers resembled bodies of water seen at a distance.

implied that there was nothing supernatural or special about the Moon or, by implication, about the planets and the stars either. Followed to its conclusion, the observation implied that there was perhaps nothing divine or extraordinary about Earth itself. Earth was one of many bodies in space, like the Moon and the other planets.

What You Can See

Even if you don't have a telescope, there are some very interesting lunar observations you can make. You might, for example, try tracking its daily motion against the background stars. Because the Moon travels around Earth (360 degrees) in 27.3 days, it will travel through about 13 degrees in 24 hours, or about half a degree (its diameter) every hour.

Have you ever thought that the Moon looks bigger when it's closer to the horizon? It's just an optical illusion, but you can test it out. The angular size of the Moon is surprisingly small. Half of your index finger, seen with your arm fully extended should cover the Moon.

The telescope through which Galileo Galilei made his remarkable lunar observations was a brand-new and very rare instrument in 1609; but you can easily surpass the quality of his observations with just binoculars.

No other celestial object is so close to us. Being so close, the Moon provides the most detailed images of an extraterrestrial geography that you will ever see through your own instrument.

When should you look at the Moon?

The easy answer is anytime the sky is reasonably clear. But if you're thinking that you should always wait until the Moon is full, think again. Rugged earthly landscapes are at their most dramatic when the Sun is low—early in the morning or late in the afternoon—and the light rakes across Earth, so that shadows are cast long and all stands in bold relief. The same holds true for the Moon. When you can see the sunrise or sunset line (the terminator), and the Moon is not so full as to be blindingly bright, *that* is when the topography of the Moon will leap out at you most vividly.

Take the time to observe the Moon through all of its phases.

On clear nights when the Moon is about three or four days "old," Mare Crisium and other vivid features—including the prominent craters Burckhardt and Geminus—become dramatically visible. You can also begin to see Mare Tranquilitatis, the Sea of Tranquility.

At day seven, the Moon is at its first quarter. At this time, mountains and craters are most dramatically visible. Indeed, this is the optimum night for looking at lunar features in their most deeply shadowed relief.

As the Moon enters its waxing gibbous phase beyond first quarter, its full, bright light is cheerful, but it's so bright that it actually becomes more difficult to make out sharp details on the lunar surface, although you do get great views of the eastern maria, the lunar plains.

Past day 14, the Moon begins to wane as the sunset terminator moves slowly across the lunar landscape. At about day 22, the Apennine Mountains are clearly visible. It was these mountains that Galileo studied most intensely, attempting to judge their height by the shadows they cast.

During the late waning phase of the Moon, moonrise comes later and later at night, as the Moon gradually catches up with the Sun in the sky. By the time the Moon passes day 26, it is nothing but a thin crescent of light present in the predawn sky. The new Moon follows, and as the Moon overtakes the Sun, the crescent reappears (on the other side of the Moon at sunset), and it begins to wax again.

Cold, Hard Facts About a Cold, Hard Place

The Moon is Earth's only natural satellite, and, as noted previously, it is a very large satellite for a planet as small as Earth. The planet Mercury is only slightly larger than the Moon. The mean distance between Earth and the Moon, as it orbits Earth from west to east, is 239,900 miles (386,239 km). The Moon is less than one third the size of Earth, with a diameter of about 2,160 miles (3,476 km) at its equator. Moreover, it is much less massive and less dense than Earth—$\frac{1}{80}$ as massive, with a density of 3.34 g/cm^3, in contrast to 5.52 g/cm^3 for Earth. Here's one way to think about relative sizes and distances: If Earth were the size of your head, the orbiting Moon would be the size of a tennis ball 30 feet away.

Because the Moon is so much less massive than Earth, and about a third as big, its surface gravity is about one sixth that of our planet. That's why the *Apollo* astronauts could skip and jump like they did, even wearing those heavy space suits. If you weigh 160 pounds on Earth's surface, you would weigh only 27 pounds on the Moon. This apparent change would give you the feeling of having great strength, because your body's muscles are accustomed to lifting and carrying six times the load that burdens

Star Words

A celestial object is in **synchronous orbit** when its period of rotation is exactly equal to its average orbital period; the Moon, in synchronous orbit of Earth, presents only one face to Earth.

them on the Moon. Of course, your mass—how much matter is in you—does not change. If your mass is 60 kilograms (kg) on Earth, it will still be 60 kg on the Moon.

The Moon is in a *synchronous orbit* around Earth; that is, it rotates once on its axis every 27.3 days, which is the same time that it takes to complete one orbit around Earth. Thus synchronized, we see only one side of the Moon.

It's a Moon!

No one can say with certainty how the Moon was formed (we weren't there!), but four major theories have been advanced. Let's look at each in turn.

Astro Byte

The Moon is much smaller in the night sky than you might think. Cut a small circle (about ⅕ inch in diameter) from a piece of paper. Hold that paper out at arm's length. That is how large the Moon is in the night sky. And don't let anyone tell you that a harvest moon (a full moon in September) is any larger than any other full moon. The Moon might be lower in the sky in autumn (as is the Sun), so it might look larger, but its size is unchanged.

A Daughter?

The oldest of the four theories speculates that the Moon was originally part of Earth, and was somehow spun off as a rapidly spinning, partially molten, newly forming planet.

Once prevalent, this theory (sometimes referred to as the fission theory) has largely been rejected, because it does not explain how the proto-Earth could have been spinning with sufficient velocity to eject the material that became the Moon. Moreover, it is highly unlikely that such an ejection would have put the Moon into a stable Earth orbit.

A Sister?

Another theory holds that the Moon formed separately near Earth from the same material that formed Earth. In effect, Earth and the Moon formed as a double-planet system.

This theory seemed quite plausible until lunar rock samples were recovered, revealing that the Moon differs from Earth not only in density, but in composition. If the two bodies had formed out of essentially the same stuff, why would their compositions be so different?

A Captive?

A third theory suggests that the Moon was formed independently and far from Earth, but was later captured by Earth's gravitational pull when it came too close.

This theory can account for the differences in composition between Earth and the Moon, but it doesn't explain how Earth could have gravitationally captured such a large Moon. Indeed, attempts to model this scenario with computer simulations have failed—the rogue moon can't be captured gravitationally unless the two collide. Moreover, although the theory accounts for some of the chemical *differences* between Earth and Moon, it does not explain the many chemical *similarities* that also exist.

A Fender Bender?

The favored theory today combines elements of the daughter theory and the capture theory in something called an impact theory. Most astronomers now believe that a very large object, roughly the size of Mars, collided with Earth when it was still molten and forming. Assuming the impact was a glancing one, it is suggested that shrapnel from Earth and the remnant of the other planetesimal (a planet in an early stage of formation) were ejected and then slowly coalesced into a stable orbit that formed the Moon.

This model is also popular because it explains some unique aspects of Earth (the "tip" of its rotational axis, for instance) and the Moon. In the impact model, it is further theorized that most of the iron core of the Mars-sized object would have been left behind on Earth, eventually to become part of Earth's core, while the material that would coalesce into the Moon retained less of this metallic component. This model therefore explains why Earth and the Moon share similar mantles (outer layers) but apparently differ in core composition.

Give and Take

Remember Newton's law of gravitation from Chapter 2? Newton proposed that every object with mass exerts a gravitational pull or force on every other object with mass in the universe. Well, Earth is much more (80 times more) massive than the Moon, which is why the Moon orbits us, and not we it. (If you want to get technical, we both actually orbit an imaginary point called the center of mass.) However, the Moon *is* sufficiently massive to make the effects of its gravitational field felt on Earth.

Anyone who lives near the ocean is familiar with tides. Coastal areas experience two high and two low tides within any 24-hour period. The difference between high and low tides is variable, but, out in the open ocean, the difference is somewhat more than three feet. If you've ever lifted a large bucket of water, you know how heavy water is. Imagine the forces required to raise the level of an entire ocean three or more feet!

What force can accomplish this?

The *tidal force* of gravity exerted by the Moon on Earth and its oceans.

The Moon and Earth mutually pull on each other; Earth's gravity keeping the Moon in its orbit, the Moon's gravity causing a small deformity in Earth's shape. This deformity results because the Moon does not pull equally on all parts of Earth. It exerts more force on parts of Earth that are closer, and less force on parts of Earth that are farther away.

Newton told us that gravitational forces decrease with the square of the distance. These differential or tidal forces are part of the cause of Earth's slightly distorted shape—it's ovoid rather than a perfect sphere—and they also make the oceans flow to two locations on Earth: directly below the Moon, and on the opposite side. This flow causes the oceans to be deeper at these two locations, which are known as the *tidal bulges.* The entire Earth is pulled by the Moon into a somewhat elongated football shape, but the oceans, being less rigid than the earth, undergo the greatest degree of deformity.

Star Words

A **tidal bulge** is any deformation of a celestial body caused by the gravitational force of another celestial body. The Moon creates an elongation of Earth and its oceans—a tidal bulge.

Interestingly, the side of Earth farthest from the Moon at any given time also exhibits a tidal bulge. This is because Earth experiences a stronger gravitational pull than the ocean on top of it, and Earth is "pulled away" from the ocean on that side. As Earth rotates (once every 24 hours) beneath the slower-orbiting Moon (once every 27.3 days), the forces exerted on the water cause high and low tides to move across the face of Earth.

The tides of largest range are the spring tides, which occur at new moon, when the Moon and the Sun are in the same direction, and at full moon, when they are in opposite directions. The tides of smallest range are the neap tides, which occur when the Sun and the Moon are at 90 degrees to one another in the sky. Tides affect you every day, of course, especially if you happen to be a sailor or a fisherman. But even if you live high and dry in Kansas or Nebraska, say, tides (and the Moon) still affect you. Every day, Earth is spinning a little slower on its axis because of the Moon.

Earth's rotation is slowing down at a rate that increases the length of a day by approximately 2 milliseconds ($^2/_{1,000}$ of a second) every century. Over millions of years, this slowing effect adds up. Five hundred million years ago, a day was a little over 21 hours long, and a year (1 orbit of the Sun) was packed with 410 days. When a planetesimal plowed into Earth early in the history of the solar system, Earth was rotating once every six hours. (And you think there aren't enough hours in the day now!)

How does this happen? Well, let's think again of why tides occur. The Moon's gravity causes two bulges to form in Earth's oceans, and Earth rotates (once every 24 hours) beneath that bulge. As Earth spins, friction between it and the oceans tends to pull the high tide ahead, so that the "bulge" actually leads the position of the Moon overhead.

With the ocean's bulge thus slightly ahead of the Moon's position, the Moon's gravity exerts a force that tends to slow rotation. Eventually, Earth's rotation will slow sufficiently to become synchronized with the orbit of the Moon around Earth. When that happens, the Moon will always be above the same point on Earth, and Earth's rotation period will have slowed (billions of years from now) from its present 24 hours to 47 days.

But that is only half of the picture. Earth can't be slowing down without something else speeding up (as a result of one of the fundamental conservation laws of physics). What's speeding up? *The Moon.* And what does this mean? That it's spiraling away slowly, and getting smaller and smaller in the sky.

Green Cheese?

Even with the naked eye, the Moon doesn't look particularly green. And Neil Armstrong confirmed that the surface of the Moon was more dusty than cheesy. On any night that the Moon is visible, the large, dark maria are also clearly visible. These are vast plains created by lava spread during a period of the Moon's evolution marked by intense volcanic activity. The lighter areas visible to the naked eye are called highlands. Generally, the highlands represent the Moon's surface layer, its *crust*, while the maria consist of much denser rock representative of the Moon's lower layer, its *mantle*. The surface rock is fine-grained, as was made dramatically apparent by the image of the first human footprint on the Moon. The mare resemble terrestrial *basalt*, created by molten mantle material that, through volcanic activity, swelled through the crust.

Star Words

The **crust** is the surface layer of a planet. The **mantle** is the layer beneath the crust and surrounding the **core**. **Basalt** is rock produced by volcanic activity.

This Place Has No Atmosphere

The mass of the Moon is insufficient for it to have held on to its atmosphere. As the Sun heated up the molecules and atoms in whatever thin atmosphere the Moon might have once had, they drifted away into space. With no atmosphere, the Moon has no weather, no erosion—other than what is caused by asteroid impacts—and no life. Although it was thought that the Moon had absolutely no water, recent robotic lunar missions have shown that there might be water (in the form of ice) in the permanent shadows of the polar craters.

This footprint was left in the lunar dust by an Apollo 11 *astronaut. Unless a stray meteoroid impact obliterates it, the print will last for millions of years on the waterless, windless Moon.*

(Image from NASA)

A Pocked Face

Look at the Moon through even the most modest of telescopes—as Galileo did—or binoculars, and you are impressed first and foremost by the *craters* that pock its surface.

Most craters are the result of asteroid and meteoroid impacts. Only about a hundred craters have been identified on Earth, but the Moon has thousands, great and small. Was the Moon just unlucky? No. Many meteoroids that approach Earth burn up in our atmosphere before they strike ground. And the traces of those that do strike the ground are gradually covered by the effects of water and wind erosion, as well as by plate tectonics. Without an atmosphere, the Moon has been vulnerable to whatever has come its way, preserving a nearly perfect record of every impact it has ever suffered.

Taken by Neil Armstrong on July 20, 1969, this is the first picture of the lunar surface made by a human standing on it.

(Image courtesy of NASA/ Johnson Space Flight Center)

Meteoroid collisions release terrific amounts of energy. Upon impact, heat is generated, melting and deforming the surface rock, while pushing rock up and out and creating an *ejecta blanket* of debris, including large boulders and dust. It is this ejected material that covers much of the lunar surface.

It is believed that the rate of meteoroid impact with the Moon (and with other objects in the solar system) was once much higher than it is now. The rate dropped sharply about 3.9 billion years ago—at the end of the period in which it is believed that the planets of the solar system were formed—and, some time later, lunar volcanic activity filled the largest craters with lava, giving many of them a smooth-floored appearance.

> **Star Words**
>
> **Crater,** the Latin word for "bowl," refers to the shape of depressions in the Moon or other celestial objects created (mostly) by meteoroid impacts. An **ejecta blanket** is the debris displaced by a meteoroid impact.

And What's Inside?

The Moon is apparently as dead geologically as it is biologically. Astronauts have left seismic instruments on the lunar surface, which have recorded only the slightest seismic activity, barely perceptible, in contrast to the exciting (and sometimes terribly destructive) seismic activity common on Earth and some other bodies in the solar system (such as Io, a moon orbiting Jupiter).

It is believed, then, that the interior of the Moon is uniformly dense, poor in heavy elements (such as iron) but high in silicates. The core of the Moon, about 250 miles (402 km) in diameter, might be partially molten. Around this core is probably an inner mantle, perhaps 300 miles (483 km) thick, consisting of semisolid rock, and around this layer, a solid outer mantle some 550 miles (885 km) thick. The lunar crust is of variable thickness, ranging from 40 to 90 miles (64 to 145 km) thick.

The Moon is responsible for everything from Earth's tides to the length of our day, and perhaps the presence of seasons. Most astronomers think that the Moon is with us today because of a gigantic collision early in the life of the solar system. The Moon's gravity pulls tides across Earth's surface, and its presence has slowed the rotation of Earth from a frenetic 6 hours to our current 24. Think of that the next time you see the Moon shining peacefully over your head.

The Least You Need to Know

- The features of the lunar surface are at their most dramatic when the Moon is in one of its crescent phases; however, there are fascinating observations to make during all lunar phases.

- Earth's gravitational field holds the Moon in orbit, but the Moon's gravitation also profoundly influences Earth, creating ocean tides and slowing our days ever so slightly over the millennia.

- The Moon is biologically dead and geologically inactive.

- There is abundant evidence that the Moon was formed as a result of a collision between Earth and another planet-size object very early in the history of the solar system.

Chapter 7

Solar System Family Snapshot

In This Chapter

- ◆ A solar system inventory
- ◆ Distances within the solar system
- ◆ Introduction to the Sun
- ◆ Planetary stats
- ◆ Terrestrial and jovian planets
- ◆ All about asteroids
- ◆ Comet hunting
- ◆ Meteors, meteoroids, and meteorites

A snapshot freezes an instant in time. When we think about our solar system, we usually assume that it has always been much as it is now, and always will be. But what we know of the solar system (4,000 years of accumulated knowledge) is only a mere snapshot in comparison to its 4.6 billion–year age. It took humankind millennia to reach the conclusion that our planet is part of a solar system, one of many planets spinning on its axis orbiting the Sun. These were centuries of wrestling with the Earth-centered planetary system first of Aristotle, then of Ptolemy, trying to make the expected planetary orbits coincide with actual observation.

This knowledge arose in some sense as a side product of the real initial goals: to be able to predict the motion of the planets and stars for the purpose of creating calendars and (in some cases) as a means of fortune telling. However, even the earliest astronomers (of whom we know) wanted to do more than predict planetary motion. They wanted to know what was "really" going on. When Copernicus, Galileo, Tycho Brahe, and Kepler finally succeeded in doing this quite well in the sixteenth and seventeenth centuries, it was a momentous time for astronomy and human understanding.

Understanding how the planets move is important, of course, but our understanding of the solar system hardly ends with such knowledge. In the last few decades of the twentieth century and now into the twenty-first, astronomers have learned more about the solar system than in all the 400 years since planetary motions were pretty well nailed down. As this chapter will show, the planetary neighborhood is a very interesting place, and our own world, Earth, is unique among the planets as a home to life.

A Beautiful Day in the Neighborhood: Let's Take a Stroll

Although it appears that about half of all stars are located in binary systems (two stars), our solar system is centered on a single star, the Sun. In orbit around the Sun are nine planets (in order of distance from the Sun): Mercury, Venus, Earth, Mars, Jupiter, Saturn, Uranus, Neptune, and Pluto. Around some of these planets orbit moons—about 90 at latest count. By the 1990s, astronomers had observed more than 6,000 large *asteroids*, of which approximately 5,000 have been assigned catalog numbers. (Such an assignment is made as soon as accurate orbital data is recorded.) Most asteroids are rather small; it is estimated that there are 1 million with diameters greater than 1 km (or about ⅗ of a mile). Some, perhaps 250, have diameters of at least 62 miles (100 km). About 30 have diameters of more than 124 miles (200 km). All of these planets and asteroids are the debris from the formation of the Sun. They coalesced slowly through the mutual attraction of gravity.

> **Star Words**
>
> An **asteroid** is one of thousands of small members of the solar system that orbit the Sun. Most asteroids are found between Mars and Jupiter in the asteroid belt. A few asteroids are so large that some astronomers call them minor planets or planetoids. Ceres, the largest known minor planet, has a diameter of about 581 miles (935 km).

In addition, the solar system contains a great many comets and billions of smaller, rock-size meteoroids.

Give a dog a bone. This image shows a computer model (based on radar observations) of the asteroid 216 Kleopatra. The asteroid is located in the asteroid belt between Mars and Jupiter, and is about 135 miles (217 km) long and about 58 miles (94 km) wide.

(Image from NASA/JPL)

Some Points of Interest

The orbits of the planets lie nearly in the same plane, except for Mercury and Pluto, which deviate from this plane by 7 degrees and 17 degrees, respectively. Between the orbit of Mars and Jupiter is a concentration of asteroids known as the asteroid belt. Most of the solar system's asteroids are here.

The orbits of the planets are not equally spaced, tending (very roughly) to double between adjacent orbits that are farther from the Sun.

To say that the distances between the planets and the Sun are very great is an understatement. Interplanetary distances are so great that it becomes awkward to speak in terms of miles or kilometers. For that reason, astronomers have agreed on something called an *astronomical unit (A.U.)*, which is the average distance between Earth and the Sun—that is, 149,603,500 kilometers or 92,754,170 miles.

Let's use these units to gauge the size of the solar system. From the Sun to the average distance of the outermost planet, Pluto, is 40 A.U. (3,710,166,800 miles, or almost 6 billion km). At just about a million times the radius of Earth, that's quite a distance. Think of it this way: If Earth were a golf ball, Pluto would be a chickpea about 8 miles away, Jupiter would be a basketball about 1 mile away, and the Sun would go floor-to-ceiling in a 10-foot room and be less than a quarter-mile away. However, compared to, say, the distance from Earth to the nearest star (after the Sun), even Pluto is a near neighbor. Forty A.U. is less than $\frac{1}{1,000}$ of a light-year, the distance light travels in one year: almost 6 trillion miles. Alpha Centauri, the nearest star system to our Sun, is about

Star Words

An **astronomical unit (A.U.)** is a unit of measurement equivalent to the average distance from Earth to the Sun (149,603,500 kilometers or 92,754,170 miles). To give a sense of scale, if Earth were a golf ball, with a diameter of somewhat more than 1½ inches (about 4 cm), the Sun would be located about 1,640 feet (500 m) away.

4.3 light-years from us (more than 25 trillion miles). On our golf ball scale, Alpha Centauri would be about 55,000 miles away. Not even Tiger Woods has a drive like that.

More or Less at the Center of It All

Near the center of the solar system—more accurately, at one focus of the elliptical orbits of the planets—is the Sun. The Sun is most of the solar system, containing more than 99.9 percent of the matter that makes up the solar system. All of Chapter 12 is devoted to the Sun, but be aware now just how massive an object it is. Jupiter, the largest planet in the solar system, is over 300 times the mass of Earth. And the Sun, in turn, is more than a *thousand* times more massive than Jupiter (and about 300,000 times more massive than Earth).

Planetary Report Card

Let's make a survey of the planets. Here's what we'll be measuring and comparing in the table that follows:

- **Semi-major axis of orbit.** Planets orbit the Sun not in perfectly circular paths, but in elliptical ones. The semi-major axis of an ellipse is the distance from the center of the ellipse to its farthest point. This distance doesn't exactly correspond to the distance from the Sun to the farthest point of a planet's orbit, because the Sun isn't at the exact center of the ellipse, but at one of the ellipse's two foci. We will express this number in A.U.

- **Sidereal period.** The time it takes a planet to complete one orbit around the Sun, usually expressed in Earth years.

- **Mass.** The quantity of matter a planet contains. The mass of Earth is 5.977×10^{24} kg. We will narcissistically assign Earth's mass the value of 1.0 and compare the masses of the other planets to it.

- **Radius.** At the equator, the radius of Earth is slightly less than 3,963 miles (6,400 km). We will assign the radius of Earth a value of 1.0 and compare the radii of other planets to it.

- **Number of known moons.** Self-explanatory—an ever-changing number for the outer planets. The parenthetical figures for Jupiter and Saturn indicate the number of small satellites that have been discovered since 2000.

- **Average density.** This value is expressed in kilograms of mass per cubic meter. The substance of the inner planets is dense and tightly packed; in the outer planets, the densities are typically lower.

Planet	Semi-Major Axis of Orbit (in A.U.)	Sidereal Period (in Years)	Mass (in Earth Masses)	Radius (in Earth Masses)	Moons	Density (kg/m³)
Mercury	0.39	0.24	0.055	0.38	0	5,400
Venus	0.72	0.62	0.81	0.95	0	5,200
Earth	1.0	1.0	1.0	1.0	1	5,500
Mars	1.5	1.9	0.11	0.53	2	3,900
Jupiter	5.2	11.9	318	11.2	17(11)	1,300
Saturn	9.5	29.5	95	9.5	18(10)	700
Uranus	19.2	84	15	4.0	21	1,200
Neptune	30.1	165	17	3.9	8	1,700
Pluto	39.5	249	0.003	0.2	1	2,300

The Inner and Outer Circles

Astronomers think of the planets as falling into two broad categories—with one planet left over. The four planets (including Earth) closest to the Sun are termed the *terrestrial planets*. The four farthest from the Sun (not counting Pluto) are the *jovian planets*. And Pluto, usually the farthest out of all, is a bit of a mongrel. Its location is jovian, but its size and composition put it more in a class with the moons of the jovian planets. Some astronomers prefer to think of it as the largest of the many miscellaneous objects in our solar system rather than as the smallest (and hardest-to-categorize) planet.

Star Words

The **terrestrial planets** are the rocky planets closest to the Sun: Mercury, Venus, Earth, and Mars. The **jovian planets** are the gaseous planets farthest from the Sun: Jupiter, Saturn, Uranus, and Neptune. Pluto is classed as neither terrestrial nor jovian, and in recent years astronomers have questioned its planetary status.

Snapshot of the Terrestrial Planets

Mercury, Venus, Mars, and Earth are called the terrestrial planets because they all possess certain Earth-like (terrestrial) properties. These include proximity to the Sun (within 1.5 A.U.), relatively closely spaced orbits, relatively small masses, relatively small radii, and high density (rocky and solid-surfaced). Compared to the larger, more distant jovian planets, the terrestrials rotate more slowly, possess weak magnetic fields, lack rings, and have few or no moons. We'll take a closer look at the terrestrial planets in Chapter 8.

Snapshot of the Jovian Planets

The jovians are far from the Sun and travel in widely spaced orbits. They are massive planets with large radii, yet they are of low density with predominantly gaseous makeup and no solid surface. In contrast to the terrestrial planets, they rotate faster, possess strong magnetic fields, have rings, and are orbited by many large moons. We'll discuss the jovian planets in Chapter 9.

Serving Up the Leftovers

What's the oldest stuff in your refrigerator (aside from that rubbery celery you bought but never ate)? Leftovers! The same is true in the solar system. The fragmentary leftovers of the formation of the Sun and planets are some of the oldest objects in the solar system. For a long time, few scientists paid much attention to this debris or knew much about it. More recently, however, astronomers have come to realize that many significant clues to the origin and early evolution of the solar system are to be found not in the planets but in the smaller bodies, the planetary moons and solar system debris. For the most part, the planets are very active places. Atmospheres have produced erosion, and internal geological activity has erased ancient surfaces. On Earth, weather, water, and tectonic motion have long since "recycled" the original surface of the planet.

So studying the planets reveals relatively little about the origins of the solar system. However, on moons and asteroids, atmospheres are sparse or nonexistent, and geological activity is minimal or absent. The result? Many of these bodies have changed little since the solar system was born. They are, in effect, cosmic leftovers.

Star Words

A **retrograde** orbit is one that is backward or contrary to the direction of the planets. An **ellipse** is called **eccentric** when it is noncircular. An ellipse with an eccentricity of 0 is a circle, and an ellipse with an eccentricity of close to 1 would be very oblong. Even planets with the most eccentric orbits (Pluto and Mercury) have eccentricities of only about 0.2.

The Asteroid Belt

Astronomers have noted and cataloged more than 6,000 asteroids with regular orbits, most of them concentrated in the asteroid belt, between the orbits of Mars and Jupiter. So far, every asteroid that has been observed orbits in the same direction as Earth and other planets—except one, whose orbit is *retrograde* (backward, or contrary to the direction of the planets). Although the asteroids move in the same direction—and pretty much on the same plane—as the planets, the shape of their orbits is different. Many asteroid orbits are more *eccentric* (the ellipse is more exaggerated and oblong) than those of the planets.

Landing on Eros—The Love Boat

In early 2001, an asteroid-exploring probe orbited and finally crash-landed on the surface of Eros, an asteroid. As it approached the asteroid's surface, it sent back tantalizing close-up images of the surface.

These images of Eros show the closest views of an asteroid ever seen. The top images are 550 m across, and the bottom images are 230 m across. The surface is dominated by small boulders.

(Image from NASA)

Rocks and Hard Places

Asteroids are composed of stony as well as metallic—mostly iron—materials and are basically tiny planets without atmospheres. Some asteroids have a good deal of carbon in their composition as well. These, called carbonaceous chondrites, are thought to represent the very first materials that came together to form the objects of the solar system. Carbonaceous chondrites are truly ancient messengers, having avoided change for billions upon billions of years.

Earlier astronomers surmised that asteroids were fragments resulting from various meteoric collisions. Although some of the smaller meteoroids were likely produced this way, the major asteroids probably came into being at the time of the formation of the solar system as a whole. Theoretical studies show that no planet could have formed at the radius of the asteroid belt (about 3 A.U. from the Sun). The region between Mars and Jupiter is dominated by the gravitational influence of the giant planet Jupiter. This force stirred up the potential planet-forming material, causing it to collide and break up instead of coming together to create a planet-sized object.

The smaller asteroids come in a wide variety of shapes, ranging from nearly spherical, to slablike, to highly irregular.

> ### Astronomer's Notebook
>
> From 1993 to 1994, the Galileo probe passed through the asteroid belt on its way to Jupiter and took pictures of an asteroid orbited by its own miniature moon. The potato-shaped asteroid named Ida is about 35 miles (56 km) long, and it is orbited at a distance of roughly 60 miles (97 km) by a rock less than 1 mile in diameter. This little moon is the smallest known natural satellite in the solar system.

Impact? The Earth-Crossing Asteroids

Most of the asteroids in the asteroid belt remain there, but some have highly eccentric orbits that take them out of the asteroid belt and across the orbital path of Earth (as well as the paths of other terrestrial planets).

Nearly 100 of these so-called *Apollo asteroids* have been identified so far, and a number of astronomers advise funding efforts to identify and track even more, because the potential for a collision with Earth is all too real—doomsday predictions about upcoming collisions or near misses make the news with fair regularity. With advance warning, some scientists believe, missiles with thermonuclear warheads could be exploded near an incoming asteroid, sufficiently altering its course so it would avoid Earth, or shattering it into a large number of smaller asteroids. Your local movie theater or video store is a good source to study Hollywood's take on these nightmare scenarios, but they are a very credible threat. Project NEAT (Near Earth Asteroid Tracking) is funded by NASA. For more information see http://neat.jpl.nasa.gov/.

It is believed that a few asteroids of more than a half-mile diameter might collide with Earth in the course of a million years. Such impacts would be disastrous, each the equivalent of the detonation of several hydrogen bombs. Not only would a great crater, some eight miles across, be formed, but an Earth-enveloping dust cloud would darken the skies. It is thought that the great extinction of dinosaurs 65 million years ago was due to such an impact. Were the impact to occur in the ocean, tidal waves and massive flooding would result. Earth impacts of smaller objects are not uncommon, but on June 30, 1908, a large object—apparently the icy nucleus of a small comet—fell in the sparsely inhabited Tunguska region of Siberia. The falling object outshone the Sun, and its explosive impact was felt at a distance of more than 600 miles.

Star Words

Apollo asteroids are those with sufficiently eccentric orbits to cross paths with Earth (and other terrestrial planets).

A very wide area of forest was obliterated—quite literally flattened. Pictures from the time show miles of forest with trees stripped and lying on their sides like matchsticks, pointing away from the impact site.

Anatomy of a Comet

The word *comet* derives from the Greek word *kome*, meaning "hair." The name describes the blurry, diaphanous appearance of a comet's long tail.

But the tail is only part of the anatomy of a comet, and it is not even a permanent part, forming as the comet nears the Sun. For most of the comet's orbit, only its main, solid body—its nucleus—exists. It is a relatively small (a few miles in diameter) mass of irregular shape made up of ice and something like soot, consisting of the same elements that we find in asteroids.

The orbit of the typical comet is extremely eccentric (elongated), so that most comets (called long-period comets) travel even beyond Pluto and might take millions of years to complete a single orbit. So-called "short-period" comets don't venture beyond Pluto and, therefore, have much shorter orbital periods.

As a comet approaches the Sun, the dust on its surface becomes hotter, and the ice below the crusty surface of the nucleus sublimates—that is, immediately changes to a gas without first becoming liquid. The gas leaves the comet, carrying with it some of the dust. The gas molecules absorb solar radiation, then reradiate it at another wavelength while the dust acts to scatter the sunlight. The effect of this is the creation of a *coma*, a spherical envelope of gas and dust (perhaps 60,000 miles across) surrounding the nucleus and a long tail consisting of gases and more dust particles.

Star Words

A **comet** is a "dirty snowball," a small celestial body composed mainly of ice and dust, which has a highly eccentric orbit around the Sun. As it approaches the Sun, some of its material is vaporized and ionized to create an atmosphere ("coma") and long tail.

A Tale of Two Tails

Most comets actually have two tails. The dust tail is usually broader and more diffuse than the ion tail, which is more linear. The ion tail is made up of ionized atoms—that is, atoms that are electrically charged. Both the dust tail and the ion tail point away from the Sun. But the dust tail is usually seen to have a curved shape that trails the direction of motion of the comet. Careful telescopic or binocular observations of nearby comets can reveal both of these tails.

What we cannot see optically is the vast hydrogen envelope that surrounds the coma and the tail. It is invisible to optical observation.

Common sense tells us that the tail would stream behind the fast-moving nucleus of the comet. This is not the case, however. The ion tail (when the comet is far from the Sun) or tails (the dust tail appears as the comet gets close to the Sun) point away from the sun, regardless of the direction of the comet's travel. Indeed, as the comet rounds the Sun and begins to leave the Sun's proximity, the tail actually *leads* the nucleus and coma. This is because the tail is "blown" like a wind sock by the *solar wind*, a constant stream of matter and radiation that escapes from the Sun. It was by observing the behavior of comet tails that astronomers discovered the existence of the solar wind.

> **Star Words**
>
> **Solar wind** is the continuous stream of radiation and matter that escapes from the Sun. Its effects are seen when it blows the tails of a comet approaching the Sun.

"Mommy, Where Do Comets Come From?"

The solar system has two cometary reservoirs, both named after the Dutch astronomers who discovered them. The nearer reservoir is called the Kuiper Belt. The short-period comets, those with orbital periods less than 200 years, are believed to come from this region, which extends from the orbit of Pluto out to several 100 A.U. Comets from this region orbit peacefully unless some gravitational influence sends one into an eccentric orbit that takes it outside of the belt.

Long-period comets, it is believed, originate in the Oort Cloud, a vast area (some 50,000–100,000 A.U. in radius) surrounding the solar system and consisting of comets orbiting in various planes. Oort comets are thought to be distributed in a spherical cloud instead of a disk.

> **Close Encounter**
>
> Comets do not randomly occur, but are regular, orbiting members of the solar system. The most famous comet of all is Halley's Comet, named after the British astronomer Edmund Halley (1656–1742), who published a book in 1705, showing by mathematical calculation that comets observed in 1531, 1607, and 1682 were actually a single comet. Halley predicted the comet would return in 1758, and, when it did, it was named in his honor. Subsequent calculations show that Halley's Comet, which appears every 76 years, had been seen as early as 240 B.C.E. and was always a source of great wonder and even fear.

The Oort Cloud is at such a great distance from the Sun, that it extends about one third of the distance to the nearest star. We don't see the vast majority of these comets, because their orbital paths, though still bound by the Sun's gravitational pull, never approach the perimeter of the solar system. However, it is believed that the gravitational field of a passing star from time to time deflects a comet out of its orbit within the Oort Cloud, sending it on a path to the inner solar system, perhaps sealing our fate.

After a short-period or long-period comet is kicked out of its Kuiper Belt or Oort Cloud home, it assumes its eccentric orbit indefinitely. That is, it can't go home again. A comet will, each time it passes close to the Sun, lose a bit of its mass as it is boiled away. A typical comet loses about $\frac{1}{1000}$ of its mass each time it passes the Sun, and so, after 100 passages, will typically fragment and continue to orbit as a collection of debris or coalesce with the Sun. As Earth passes through the orbital paths of such debris, we experience meteor showers.

A-Hunting We Will Go

Visitations by major comets, such as Comet Hyakutake in 1966 and Hale-Bopp in 1997, are newsworthy events. Turn on the television or read a newspaper, and you'll be told where to look and when. But most comets don't make the front pages. For the latest comet news, check out the NASA comet home page at http://encke.jpl.nasa.gov. *Sky and Telescope* magazine also publishes comet information (see Appendix E).

Of course, you don't have to limit yourself to looking for already known comets. You can head out and hunt for new ones.

Comet hunting can be done with or without a telescope, but a good telescope greatly increases your chances of finding a new comet. Remember that telescopes catch more light than our eyes, and most comets are discovered as a tiny, wispy smudge. The coma will not appear much different from a star, but you should see a gradual, not sharply defined, tail attached to it. The tail might be a short, broad wedge or a long ion streamer.

> ### Astronomer's Notebook
>
> What are your chances of finding a new comet? David H. Levy, discoverer of several comets and co-discoverer of the celebrated Comet Shoemaker-Levy, reports that, on average, observers hunt about 400 hours for their first comet and 200 for each additional one. Apparently, comet discoveries are subject to a bulk discount.

The following tips will increase your chances of finding a comet:

♦ Choose a moonless night and go to a rural area, away from city lights.

♦ According to David H. Levy, just before dawn, two days before or five days after the new moon, is an ideal time to search.

♦ Comets can be seen in any part of the sky, but they are brightest when they approach within 90 degrees of the Sun; therefore, at sunset, look from directly overhead to the western horizon.

♦ Discovering a comet requires you to see something unusual or different in the sky. For this reason, you would do well to spend time becoming familiar with the sky so that you will be better able to recognize something new.

Catch a Falling Star

Although sighting a comet—especially a hitherto unrecorded one—is certainly a momentous event, few astronomical phenomena are more thrilling than the sight of a meteor. Best of all, such sightings are common and require no telescope.

Meteors, Meteoroids, and Meteorites

Meteors are often called shooting stars, although they have nothing to do with stars at all. A *meteor* is a streak of light in the sky resulting from the intense heating of a narrow channel in Earth's upper atmosphere. The heat generated by friction with air molecules as the meteoroid hurtles through Earth's atmosphere ionizes—stripping electrons away from atoms—a pathway behind this piece of space debris. The ionized path in Earth's atmosphere glows for a brief time producing the meteor.

Although smaller meteoroids (often called *micrometeoroids*) are typically the rocky fragments left over from a broken-up comet, the meteor phenomenon is very different from a comet. A meteor sighting is a momentary event. The meteor streaks across a part of the sky. As you read previously, a comet does not streak rapidly and might, in fact, be visible for many months because of its great distance from Earth. A meteor is an event occurring in Earth's upper atmosphere, whereas a comet is typically many A.U. distant from Earth.

Star Words

Meteor is the term for what you see streaking across the night sky—a "shooting star." **Meteoroid** is the object itself, a rocky thing that is typically a fragment shed from a comet or an asteroid. A **micrometeoroid** is a very small meteoroid. Those few meteoroids that are not consumed in Earth's atmosphere reach the ground as **meteorites**. Many science museums have collections of meteorites.

Meteor is the term for the sight of the streak of light caused by a *meteoroid*—which is the term for the actual rocky object that enters the atmosphere. Most meteoroids are completely burned up in our atmosphere, but a few do get through to strike Earth. Any fragments recovered are called *meteorites*.

Although most of the meteors we see are caused by small meteoroids associated with comet fragments (about the size of a pea), larger meteoroids, more than an inch or so, are probably asteroid fragments that have strayed from their orbit in the asteroid belt. Such fragments enter Earth's atmosphere at supersonic speeds of several miles per second and often generate sonic booms. If you see a very bright meteor—the brightness of the planet Venus or even brighter—it is one of these so-called *fireballs*. It is estimated that about 100 tons of meteoric material fall on Earth each day.

Star Words

A **fireball** results from a substantial meteoroid (measuring perhaps an inch or so in diameter), which creates a spectacular display when it enters the atmosphere. Meteorites must be rather large (larger than your head) to avoid burning up entirely before hitting the surface of Earth.

Meteor crater in Arizona was formed about 50,000 years ago as the result of the impact of an object some 80 feet in diameter. The crater is nearly a mile (over a kilometer) in diameter.

(Image from JPL/NASA)

News from NEAT

Very rarely—perhaps once in several hundred thousand years—a large meteoroid strikes Earth, causing great damage.

Counts of near-Earth aster-oids from Project NEAT (Near Earth Asteroid Tracking).

(Table from NASA/NEAT)

April Showers (or the Lyrids)

Whenever a comet makes its nearest approach to the Sun, some pieces break off from its nucleus. The larger fragments take up orbits near the parent comet, but some fall behind, so that the comet's path is eventually filled with these tiny micrometeoroids. Periodically, Earth's orbit intersects with a cluster of such micrometeoroids, resulting in a *meteor shower* as the fragments burn up in our upper atmosphere.

Meteor showers associated with certain comets occur with high regularity. They are named after the constellation from which their streaks appear to radiate. The follow-ing table lists the most common and prominent showers. The shower names are geni-tive forms of the constellation name; for example, the Perseid shower comes from the direction of the con-stellation Perseus, the Lyrids from Lyra. The dates listed are those of maximum expected activity, and you can judge the intensity of the shower by the esti-mated hourly count. The table also lists the parent comet, if known. Consult a monthly astronomy mag-azine such as *Astronomy* or *Sky & Telescope* to obtain specific peak times for a given year's meteor showers.

Star Words

When Earth's orbit inter-sects the debris that litter the path of a comet, we behold a **meteor shower,** a period when we see more meteors than the average.

Name of Shower	Maximum Activity	Estimated Hourly Count	Parent Comet
Quadrantid	January 3	50	unknown
Beta Taurid	June 30	25	Encke
Perseid	August 12	50+	1862III (Swift-Tuttle)
Draconid	October 8–9	500+	Giacobini-Zimmer
Orionid	October 20	25	Halley
Leonid	November 16–17	10*	1866I (Tuttle)
Geminid	December 11–17	50–75	3200 Phaeton

Every 33 years, Earth's orbit intersects the densest part of the Leonid debris path, resulting in the potential for a meteor infall rate of 1,000 a minute! Such an intersection occurred in 1999, and will happen again in 2032.

You'll recall from Chapter 5 that you can detect meteor showers on your FM radio or even on unused VHF television frequencies. But if it's clear outside, we suggest that you take your radio outside, and as you listen for distant radio stations to pop up, look up at the skies and watch as well. It might be hard to believe that most of those streaks of light are following meteoroids no larger than a pea. But be thankful that they are!

The Least You Need to Know

♦ Eight of the nine planets of the solar system are divided into the rocky terrestrial planets (those nearest the Sun: Mercury, Venus, Earth, and Mars) and gaseous jovian planets (those farthest from the Sun: Jupiter, Saturn, Uranus, and Neptune). Pluto, usually the most distant planet from the Sun, is not categorized, though it has a lot in common with the moons of the outer planets.

♦ Although the Sun and planets are certainly the major objects in the solar system, astronomers also pay close attention to the minor bodies—asteroids, comets, meteors, and planetary moons—that can tell us a lot about the origin of the solar system.

♦ Although most asteroids are restricted to highly predictable orbits, a few cross Earth's orbital path, posing a potentially catastrophic hazard.

♦ Comets and meteor showers present ample opportunities for exciting amateur observation.

Chapter 8

So Close and Yet So Far: The Inner Planets

In This Chapter

- ◆ Where do the planets' names come from?
- ◆ Some vital statistics
- ◆ Orbit and rotation
- ◆ Atmospheres on the terrestrial planets
- ◆ Planetary surfaces
- ◆ The geology of the terrestrials
- ◆ How everything might have been different

Our two closest neighbors in the solar system, Mars and Venus, are a constant reminder of how easily things could have turned out differently here on Earth. Venus is so hot and forbidding that it might be a good place to film Dante's *Inferno*, and although pictures of Mars might resemble the American Southwest, it has an atmosphere so cold and thin that it's hardly there at all. Putting a planet a little closer to the Sun or a little farther away can truly make all the difference in the world. Equally amazing is that all three planets fall within what is called the "habitable zone" of the

Sun (see Chapter 22). The habitable zone is the distance range from the Sun within which water can exist as a liquid on the planet's surface. But only one planet, Earth, has abundant liquid water. Recent imaging of the surface of Mars indicates that water might still exist there in liquid form, albeit only fleetingly.

In this chapter, we will take a closer look at the four rocky planets that are closest to the Sun. As noted in Chapter 7, astronomers call these four planets terrestrial, meaning Earthlike. To be sure, Mercury, Venus, and Mars resemble Earth more than they do the jovian planets—Jupiter, Saturn, Uranus, and Neptune—but, as you'll see, these four worlds are as different from each other as one could imagine. And Earth is the only one of the four that has sustained life.

Why is that? What is it about Earth that allowed life to thrive here and not on the other three planets in the inner solar system? Let's start looking for some answers.

The Terrestrial Roster

The terrestrial planets are Mercury, Venus, Earth, and Mars. Except for Earth, all are named after Roman gods. Mercury, the winged-foot messenger of the gods, is an apt name for the planet closest to the Sun; its sidereal period (the time it takes a planet to complete on orbit around the Sun) is a mere 88 Earth days, and its average orbital speed (30 miles per second or 48 km/s) is the fastest of all the planets. Mercury orbits the Sun in less than three months, or about four times for each Earth orbit.

Venus, named for the Roman goddess of love and fertility, is (to observers on Earth) the brightest of the planets, and, even to the naked eye, quite beautiful to behold. Its atmosphere, we shall see, is not so lovely. The planet is completely enveloped by carbon dioxide and thick clouds that consist mostly of sulfuric acid.

The name of the bloody Roman war god, Mars, suits the orange-red face of our nearest planetary neighbor—the planet that has most intrigued observers and that seems, at first glance, the least alien of all our fellow travelers around the Sun.

You looked at some vital statistics of the planets in Chapter 7. Here are some more numbers, specifically for the terrestrial planets. Notice that the presence of an atmosphere (for Venus and Earth) causes there to be much less variation in surface temperature.

Planet	Mass in Kilograms	Radius in Miles (and km)	Surface Gravity (Relative to Earth)	Rotation Period in Solar Days	Surface Temperature in K
Mercury	3.3×10^{23}	1,488 (2,400)	0.4	59	100–700
Venus	4.9×10^{24}	3,782 (6,100)	0.9	–243*	~730
Earth	6.0×10^{24}	3,968 (6,400)	1.0	1.0	~290
Mars	6.4×10^{23}	2,108 (3,400)	0.4	1.0	100–250

The rotation period is negative because the rotation of Venus is retrograde; that is, the planet rotates on its axis in the opposite direction from the other planets. Viewed from above the North Pole of Earth, all of the planets except Venus rotate counterclockwise. That means that on Venus, the Sun would rise (if you could see it through the thick cloud cover) in the west.

When we discuss the formation of the solar system in Chapter 11 we will mention a few observational facts that "constrain" our models of planetary system formation. As you'll see, a few rules of planetary motion are immediately apparent. All four terrestrial planets orbit the Sun in the same direction. All except Venus rotate on their axes in the same direction as they orbit the Sun. The orbital paths of the inner four planets are *nearly* circular. And the planets all orbit the Sun in roughly the same plane.

But the solar system is a dynamic and real system, not a theoretical construct, and there are interesting exceptions to these rules. The exceptions can give us insight into the formation of the solar system.

Mercury: The Moon's Twin

In many ways, Mercury has more in common with the lifeless Moon of our own planet than with the other terrestrial planets. Its face is scarred with ancient craters, the result of massive bombardment that occurred early in the solar system's history. These craters remain untouched because Mercury has no water, erosion, or atmosphere to erase them. The closest planet to the Sun—with an average distance of 960,000 miles (1,546,000 km)—Mercury is difficult to observe from Earth, and can only be viewed near sunrise or sunset.

Astro Byte

Mercury's temperature range of 600 K (it varies from 100 K to 700 K) is the greatest of all the planets. Although Mercury is closest to the Sun, Venus's surface temperature is higher. A planet's surface temperature and its mass determine how much (if any) of its atmosphere it will be able to hold on to.

Close Encounter

Most of us in the United States are accustomed to the Fahrenheit temperature scale. The rest of the world uses the Celsius (Centigrade) scale. Astronomers, like most scientists, measure temperature on the Kelvin scale. Throughout this book, we have expressed distance in the units familiar to most of our readers: miles (with kilometers or meters given parenthetically). For mass we give all values in kilograms. The Kelvin scale for temperature is conventional and very useful in astronomy; let us explain.

The Fahrenheit scale is really quite arbitrary, because its zero point is based on the temperature at which alcohol freezes. What's fundamental about that? Worse, it puts at peculiar points the benchmarks that most of us *do* care about. For example, water freezes at 32° F at atmospheric pressure and boils at 212° F. The Celsius scale is somewhat less arbitrary, because it is laid out such that water freezes at 0° C and boils at 100° C at atmospheric pressure. But the atmospheric pressure of Earth is by no means a universal quantity, so astronomers and others looked for more fundamental benchmarks.

The Kelvin scale is least arbitrary of all. It forces us to ask a fundamental question: What is heat?

The atoms and molecules in any matter are in constant random motion, which represents thermal energy. As long as there is atomic or molecular motion, there is heat (even in objects that, to the human senses, feel very cold). We know of no matter in the universe whose atoms and molecules are entirely motionless, but, in theory, such an absolute zero point does exist. The Kelvin scale begins at that theoretical absolute zero, the point at which there is no atomic or molecular motion. On the Fahrenheit scale, that temperature is −459 degrees. On the Celsius scale, it is −273 degrees. On the Kelvin scale, it is merely 0 degrees. Thus, in the Kelvin scale there are no negative temperatures, because absolute zero is—well—absolute. After all, atoms and molecules cannot get any *more* motionless than motionless, and being absolutely still would cause the constituents of atoms to violate fundamental physical principles.

Note that water freezes at 273 K and boils at 373 K. If you want to convert Kelvin temperatures to Celsius temperatures, subtract 273 (to be precise, 273.15) from the Kelvin temperature. If you want to, you can then convert the Celsius temperature to

Mercury's surface, revealed in detail for the first time in images transmitted by such unmanned probes as *Mariner 10* (in the 1970s), is pocked with craters. *Mariner 10* also discovered a weak magnetic field. As a result, astronomers concluded that the planet must have a core rich in molten iron. This contention is consistent with the planet's position closest to the center of the solar system, where most of the preplanetary matter—the seeds that formed the planets—would have been metallic in composition.

Mosaic of Mercury's surface taken by Mariner 10 *during its approach on March 29, 1974. The spacecraft was about 124,000 miles (200,000 km) above the planet. Note how closely the surface of Mercury resembles the Moon's.*

(Image from JPL/NASA)

Lashed to the Sun

In the days before space-based telescopes and probes, earthbound astronomers did the best they could to gauge the rotation of Mercury. The nineteenth-century astronomer Giovanni Schiaparelli observed the movement of what few, indistinct surface features he could discern and concluded that, unlike any other planet, Mercury's rotation was synchronous with its orbit around the Sun.

Synchronous orbit means that Mercury always keeps one face toward the Sun and the other away from it, much as the Moon always presents the same face to Earth.

Technology marches on. In 1965, by means of radar imaging (unavailable to Schiaparelli in the nineteenth century), astronomers discovered that Mercury's rotation period was not 88 days, as it was long thought to be, but only 59 days. This discovery implied that Mercury's rotation was not precisely synchronous with its orbit, but that it rotated three times around its axis for every two orbits of the Sun.

"I Can't Breathe!"

Like Earth's Moon, Mercury possesses insufficient mass to hold—by gravitation—an atmosphere for very long. In the same way that mass-attracting mass built up planetesimals, so the early planets built up atmospheres by hanging on to them with their gravitational pull. If an atmosphere was ever associated with Mercury, the heating of the Sun and the planet's small mass helped it to escape long ago. Without any atmosphere to speak of, the planet is vulnerable to bombardment by meteoroids, x-rays, and ultraviolet radiation, as well as extremes of heat and cold.

Despite the absence of atmosphere, regions at the poles of Mercury remain permanently in shadow, with temperatures as low as 125 K. These regions, and similar regions on Earth's Moon, might have retained some water in the form of ice.

Forecast for Venus: "Hot, Overcast, and Dense"

Venus's thick atmosphere and its proximity to the Sun are a cruel combination. The planet absorbs more of the Sun's energy (being closer to the Sun than Earth) and because of its heavy cloud cover, is unable to radiate away much of the heat. Even before astronomers saw pictures of the planet's surface, they knew that it would not be a welcoming place.

Until the advent of radar imaging aboard space probes such as *Pioneer Venus* (in the late 1970s) and *Magellan* (in the mid-1990s), details about the surface of Venus were a shrouded mystery. Optical photons reflect off the upper cloud layers of the planet and all astronomers can see with even the best optical telescopes is the planet's swirling upper atmosphere. Modern radio imaging techniques (which involve bouncing radio signals off the surface, as described in Chapter 5) have revealed a surface of rolling plains punctuated by a pair of raised land masses that resemble Earth's continents. Venus has no coastlines, all of its surface water having long ago evaporated in the ghastly heat. The two land masses, called Ishtar Terra and Aphrodite Terra, are high plateaus in a harsh, waterless world.

The landscape of Venus sports some low mountains and volcanoes. Volcanic activity on the surface has produced *calderas* (volcanic craters) and *coronae*, which are vast, rough, circular areas created by titanic volcanic upswellings of the mantle.

Star Words _____

Calderas are craters produced not by meteoroid impact, but by volcanic activity. **Coronae** are another effect of volcanic activity: large upswellings in the mantle, which take the form of concentric fissures, or stretch marks in the planet's surface.

Venus is surely lifeless biologically, but geologically it is very lively. Volcanic activity is thought to be ongoing, and many astronomers believe that the significant, but fluctuating, level of sulfur dioxide above the Venusian cloud cover might be the result of volcanic eruptions. Probes sent to Venus thus far have not detected a *magnetosphere;* however, astronomers believe that the planet has an iron-rich core. Scientists reason that the core of Venus might simply rotate too slowly to generate a detectable magnetic field.

Star Words

A **magnetosphere** is a zone of electrically charged particles trapped by a planet's magnetic field. The magnetosphere lies above the planet's atmosphere.

Radar mapping carried out by the Magellan *probe made this image of volcanic domes in the Eistla region of Venus in 1991.*

(Image from JPL/NASA)

The Sun Also Sets (in the East)

As we've seen, Mercury's peculiar rotational pattern can be explained by its proximity to the Sun. But no such gravitational explanation is available for the peculiar behavior of Venus. If at 59 days, Mercury rotates on its axis slowly, Venus is even more sluggish, consuming 243 Earth days to accomplish a single spin.

What's more, it spins backward! That is, viewed from a perspective above Earth's North Pole, all of the planets (terrestrial and jovian) spin counterclockwise—except for Venus, which spins clockwise.

Nobody knows why for sure, but we can surmise that the rotational peculiarities of Venus were caused by some random event that occurred during the formation of the solar system—a collision or close encounter with another planetesimal, perhaps. A violent collision, like the one that formed Earth's Moon, might have started Venus on its slow backward spin.

Venusian Atmosphere

Chemically, the atmosphere of Venus consists mostly of carbon dioxide (96.5 percent). The remainder is mostly nitrogen. These are organic gases, which might lead one to jump to the conclusion that life—*some* form of life—might exist on Venus. Indeed, during the 1930s, spectroscopic studies of Venus revealed the temperature of the planet's upper atmosphere to be about 240 K—close to Earth's *surface* temperature of 290 K. Some speculated that the environment of Venus might be a dense jungle.

In the 1950s, radio waves were used for the first time to penetrate the dense cloud layer that envelops Venus. It turned out that surface temperatures were not 240 K, but closer to 600 K. Those temperatures are incompatible with any form of life. And the outlook got only worse from there. Spacecraft probes soon revealed that the dense atmosphere of Venus creates high surface pressure—the crushing equivalent of 90 Earth atmospheres—and that surface temperatures can top 730 K.

And what about those clouds?

On Earth, clouds are composed of water vapor. But Venus shows little sign of water. Its clouds consist of sulfuric acid droplets. Talk about acid rain.

Even hot planets have a north pole. This view of Venus from the radar-mapping Magellan *space probe shows the planet seen from its north pole, a view hidden from optical view by dense swirling clouds. Bright spots are high, and dark spots are low.*

(Image from NASA)

The Earth: Just Right

In our march through the terrestrial planets, the next logical stop would be Earth. We have mentioned some of the unique aspects of our home planet in earlier chapters, and will mention more in the course of the book. In particular, we will look at Earth as a home to life when we discuss the search for life elsewhere in the Milky Way (Chapter 22).

But let's take a brief moment to think of Earth as just another one of the terrestrial planets. Earth is almost the same size as Venus, and has a rotational period and inclination on its axis almost identical to Mars. How is it, then, that Earth is apparently the only one of these three planets to support life?

As in real estate, it all comes down to three things: location, location, and location. Earth is far enough from the Sun that it has not experienced the runaway greenhouse effect of Venus. It is close enough to the Sun to maintain a surface temperature that allows for liquid water, and massive enough to hold on to its atmosphere. The molten rock in the mantle layer above its core keeps the crust of Earth in motion (called plate tectonics), and the rotation of this charged material has generated a magnetic field that protects Earth from the brunt of the solar wind.

These conditions have created an environment in which life has gotten a foothold and flourished. And life has acquired enough diversity that the occasional setback (like the asteroid that struck Earth some 65 million years ago) might change the course of evolution of life on the planet, but has not yet wiped it out.

Our home planet is truly remarkable and remarkably balanced. The more we learn about our terrestrial neighbors, the more we should appreciate the delicate balance that supports life on Earth.

Mars: "That Looks Like New Mexico!"

Those of us who were glued to our television sets in 1997 when NASA shared images of the Martian surface produced by the *Mars Pathfinder* probe were struck by the resemblance of the landscape to Earth. Even the vivid red coloring of the rocky soil seemed familiar to anyone who has been to parts of Australia or even the state of Georgia—though the general landscape, apart from its color, more closely resembles desert New Mexico. You can check out this resemblance for yourself by pointing your Web browser to NASA's *Mars Pathfinder* pages at http://mars.jpl.nasa.gov/MPF/index0.html.

In contrast to Mercury and Venus, which are barely inclined on their axes (in fact, their axes are almost perpendicular to their orbital planes), Mars is inclined at an angle of 25.2 degrees—quite close to Earth's inclination of 23.5 degrees.

And that's only one similarity. Although Mercury and Venus move in ways very different from Earth, Mars moves through space in a way that should seem quite familiar to us. It rotates on its axis once in every 24.6 hours—a little more than an Earth day—and because it is inclined much as Earth is, it also experiences familiar seasonal cycles.

The peculiarities of Mercury and Venus make Mars look more similar to Earth than it really is. Generations have looked to the red planet as a kind of solar system sibling, partly believing, partly wishing, and partly fearing that life might be found there. But the fact is that life as it exists on Earth cannot exist on the other terrestrial planets.

Martian Weather Report: Cold and Thin Skies

Although the atmosphere of Venus is very thick, that of Mars is very thin. It consists of about 95 percent carbon dioxide, 3 percent nitrogen, 2 percent argon, and trace amounts of oxygen, carbon monoxide, and water vapor. Although our imaginations might tend to paint Mars as a hot desert planet, it is actually a very cold, very dry, place—some 50 K (on average) colder than Earth.

Section of a geometrically improved, color-enhanced version of a 360-degree pan taken by the Imager for Mars Pathfinder *during three Martian days in 1997. Note the* Sojourner *rover vehicle and its tracks.*

(Image from NASA)

The Martian Chronicles

Percival Lowell was born in Boston in 1855, the son of one of New England's wealthiest and most distinguished families. His early career was absorbed in literature (his sister Amy Lowell became a famous poet) and Far Eastern travel. He became a diplomat, serving as counselor and foreign secretary to the Korean Special Mission to the United States. But in the 1890s, he read a translation of an 1877 book by Giovanni

Schiaparelli, the same Italian astronomer who had concluded that Mercury's rotation was synchronized with its orbit. Reporting his observations of the surface of Mars, Schiaparelli mentioned having discovered *canali*. The word, which means nothing more than "channels" in Italian, was translated as "canals" in the translation Lowell read, and the budding astronomer, already charmed by exotic places, set off in quest of the *most* exotic of all: Mars— and whatever race of beings had excavated canals upon it.

Astro Byte

Atmospheric pressure on Mars is only about $\frac{1}{150}$ that of Earth. On Venus the atmospheric surface pressure is about 100 times greater. Again: Earth is "just right" for us.

Lowell dedicated his considerable family fortune to the study of the planet Mars. He built a private observatory in Flagstaff, Arizona, and, after years of observation, published *Mars and Its Canals* in 1906. Noting that the canal network underwent seasonal changes, growing darker in the summer, Lowell theorized that technologically sophisticated beings had created the canals to transport crop-irrigation water from the Martian polar ice caps. In 1924, astronomers searched for radio signals from the planet (using a technique that anticipated the current SETI search for radio signals from the universe), but to no avail. Yet the idea of intelligent life on Mars was so ingrained in the public imagination that, on October 30, 1938, Orson Welles's celebrated radio adaptation of H. G. Wells's 1898 science fiction novel about an invasion from Mars, *War of the Worlds*, triggered regional panic.

A variety of space probes have now yielded very high-resolution images of Mars, revealing the apparent canals as simply natural features, such as craters and canyons. Although it is true that Mars undergoes seasonal changes, the ice caps consist of a combination of frozen carbon dioxide and water.

Why Mars Is Red

If we feel any disappointment at the loss of the Martian canals, at least we can still enjoy the image of the "angry red planet." Yet the source of the reddish hue is not the bloody spirit of the Roman god of war, but simple iron. The Martian surface contains large amounts of iron oxide, red and rusting. As *Viking 1* and *Mars Pathfinder* images revealed, even the Martian sky takes on a rust-pink tinge during seasonal dust storms.

The dust is blown about by winds that kick up in the Martian summer. These winds play a prominent role on Mars, forming vast dunes and streaking craters. An especially large dune is found surrounding the north polar cap.

Volcanoes, Craters, and a "Grand Canyon"

The *Mariner* series of planetary probes launched in the 1960s and 1970s revealed a startling difference between the southern and northern hemispheres of Mars. The southern hemisphere is far more cratered than the northern hemisphere, which is covered with wind-blown material as well as volcanic lava. There have even been proposals that the smooth northern hemisphere hides a large frozen ocean.

Volcanoes and lava plains from ancient volcanic activity abound on Mars. Because the planet's surface gravity is low (0.38 that of Earth), volcanoes can rise to spectacular heights. Like Venus, Mars lacks a strong magnetic field, but, in contrast to Venus, it rotates rapidly; therefore, astronomers conclude that the core of Mars is nonmetallic, nonliquid, or both. Astronomers believe that the core of the smaller Mars has cooled and is likely solid, consisting largely of iron sulfide.

Astro Byte

Olympus Mons, found on Mars, is the largest known volcano in the solar system. It is 340 miles (544 km) in diameter and almost 17 miles (27 km) high.

Unlike Earth, Mars failed to develop much *tectonic activity* (instability of the crust), probably because its smaller size meant that the outer layers of the planet cooled rapidly. Volcanic activity was probably quite intense some 2 billion years ago.

Close Encounter

Although you can view Mars on many nights, it is best seen for a span of a few months every 26 months when it is in opposition, meaning closest to Earth *and* on the opposite side of Earth from the Sun (that is, when Earth comes between it and the Sun). Consult one of the annual guides listed in Appendix E for the dates of the next several oppositions. When Mars is at opposition, you can make out remarkable detail, even with a modest telescope, including at least one of the polar caps as well as dark areas (maria or "seas") and contrastingly bright "desert" stretches. Colored filters can help enhance certain surface features such as the polar caps.

At 5:51 A.M. EDT on August 27, 2003, Earth passed within 34,646,418 miles (55,758,006 kilometers) of Mars, the closest approach to our planet in 59,619 years. If you missed this opposition, relax! Mars will be even closer—33,922,722 miles (54,593,328 kilometers)—on January 15, 9943.

Also impressive are Martian canyons, including Valles Marineris, the "Mariner Valley," which runs some 2,500 miles (4,025 km) along the Martian equator and is as much as 75 miles (120 km) wide and, in some places, more than four miles (6.5 km) deep. The Valles Marineris is not a canyon in the earthly sense, because it was not cut by flowing water, but is a geological fault feature.

Water, Water *Anywhere?*

Clearly visible on images produced by Martian probes are runoff and outflow channels, which are believed to be dry riverbeds, evidence that water once flowed as a liquid on Mars. Geological evidence dates the Martian highlands to four billion years ago, the time in which water was apparently sufficiently plentiful to cause widespread flooding. Recent theories suggest that at the time, Mars had a thicker atmosphere that allowed water to exist in a liquid state, even at its low surface temperatures.

The *Mars Global Surveyor* (MGS) mission has found further geological evidence for the presence of liquid and subsurface water. Such evidence has kept alive hopes that life might have existed—or might even yet exist—on a microbial level—on Mars. For the most recent images and news, check out the official MGS web site at http://mars.jpl.nasa.gov/mgs/.

All Bets Are Off

British bookies no longer take bets on whether life ever existed on Mars. The odds were 1,000:1 in the 1970s and 16:1 by the start of 2004. But the results of the latest NASA mission to Mars, reported in March 2004, clearly indicated the presence of water at some time in the history of Mars and not only removed this bet from the books, but transformed our view of the red planet as well as the solar system. The presence of water suggests the very real possibility of life, and it might well be that we can no longer think of Earth as the only inhabited real estate in the neighborhood of the Sun.

One of the primary scientific goals of the two Mars Exploration Rovers (MERs) was to determine whether or not Mars was a wetter place in the distant past. Astronomers hoped also to determine if it had been wet for extended periods of time, long enough so life might have arisen there.

In March 2004, NASA released results from the *Opportunity* rover that landed in Merdiani Planum, a region known to be rich in hematite, a form of iron that typically forms in watery environments. The presence of hematite was confirmed by the rover, but that was just a start.

When the cameras sent the first images back from the landing site, JPL scientists were amazed to see that the craft had rolled to a stop in a tiny crater—a "hole in one," as some of them described it. No ordinary crater, this one contained an outcropping of rock that, to scientists, looked like something no one had ever seen before. It was Martian bedrock.

Gone were the red stone-scattered fields of the *Viking* and *Pathfinder* missions. This was an alien landscape within an alien landscape. The surface around *Opportunity* was dark and smooth, with a clear outcropping of rock at the lip of the crater. After surveying its immediate surroundings, *Opportunity* began to examine the bedrock in detail. In it, the rover's Moessbauer spectrometer detected a hydrated (water-bearing) iron sulfate mineral called jarosite. On Earth, jarosite can form in hot, acidic environments, much like hot springs.

However, scientists require multiple lines of evidence when making a dramatic claim, and the long-term presence of water on Mars is dramatic. The other evidence for the presence of water includes indentations in the rock, spherules, and crossbedding. The indentations (called "vugs") are probably due to the formation of crystals in the rock when it sat for long periods in a salt-water environment. The spherules detected (dubbed "blueberries") are found at many layers in the rock, indicating that they might have formed when minerals seeped out of porous rock. Finally, tiny ridge patterns in the rock indicate either water or wind action over long periods of time. The small scale of the crossbedding in the outcrop rocks has geologists thinking that water is the more likely cause.

If Mars was watery for a long period of time, as these results strongly indicate, the odds for life having existed at least in one other place in the universe—indeed, in our very own solar system—go way up. For the most recent information about the Mars Exploration rovers, go to www.jpl.nasa.gov/mer2004/index.html.

Comparison between similar geological features associated with water runoff on Mars (left) and Earth (right).

(Image from JPL/NASA)

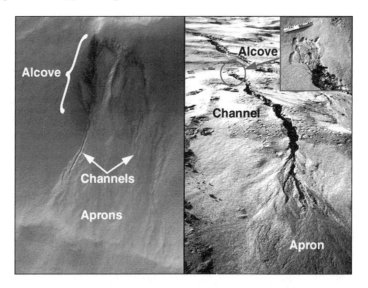

Martian Moons

Mars and Earth are the only terrestrial planets with moons. As we have said, the Moon is remarkably large, comparable in size to some of the moons of Jupiter. The moons of Mars, colorfully named Phobos (Fear) and Deimos (Panic), after the horses that drew the chariot of the Roman war god, were not discovered until 1877.

They are rather unimpressive as moons go, resembling large asteroids. They are small and irregularly shaped (Phobos is 17.4 miles long × 12.4 miles [28 km × 20 km] wide, and Deimos is 10 miles × 6.2 miles [16 km × 10 km]). They are almost certainly asteroids that were gravitationally captured by the planet and fell into orbit around it.

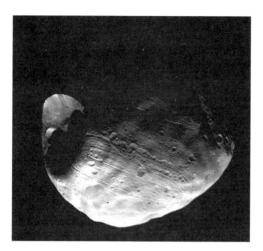

The surface of Phobos, the larger of Mars's two moons, is dominated by large craters.

(Image from JPL/NASA)

Where to Next?

Mercury, Venus, and Mars are our neighbors, the planets we know most about and, perhaps, feel the closest connection to. At that, however, they are still strange and inhospitable worlds. Strange? Inhospitable? Well, as Al Jolson was famous for saying, "You ain't seen nothin' yet." In the next chapter we visit the jovian planets, the gas giants of the outer solar system.

The Least You Need to Know

- The terrestrial planets are Mercury, Venus, Earth, and Mars.

- Although the terrestrial planets share certain Earthlike qualities, they differ in significant ways that, among other things, make the existence of life on those planets impossible or at least highly unlikely.

- Mercury and Venus display rotational peculiarities. In the case of Mercury, its rotation can be explained by its proximity to the Sun; but the slow retrograde rotation of Venus can be best explained by the occurrence of some random event (probably a collision) early in the formation of the solar system.

◆ Of the terrestrial planets, only Earth has an atmosphere and environment conducive to life.

◆ Although Mars might look and seem familiar, its thin, icy atmosphere whipped by dust storms is a harsh environment.

Chapter 9

Great Balls of Gas! The Outer Planets

In This Chapter

- ◆ The jovian planetary line-up
- ◆ Why the jovians are so gassy
- ◆ The discovery of Uranus and Neptune
- ◆ Jovian rotational quirks
- ◆ Jovian atmosphere and weather
- ◆ Interior structure of the outer giants

If you've ever been outside late at night looking to the south, chances are you've already seen the largest planet in the solar system, Jupiter. You might have thought that it was just a bright star, and that is pretty much what the ancients thought, except that they realized it moved (wandered) in a way unlike the fixed stars. Imagine Galileo's surprise, then, in 1610, when he pointed a telescope at the planet and saw its surface—it was more than a featureless point of light—and four smaller bodies orbiting it. His discovery would cause a good deal of upheaval in the way humans viewed themselves in the universe, and Galileo himself would end up in trouble with the Catholic Church. All of this because of that wandering "star" in the sky.

The planets are found near an imaginary arc across the sky that we call the *ecliptic*. Long before astronomers knew that the terrestrial planets shared common features, they knew that two of the "wanderers" that they watched were different. Although Mercury and Venus never strayed far from the Sun, and Mars moved relatively rapidly across the sky, Jupiter and Saturn moved ponderously, majestically through the stellar ocean. Long before the invention of the telescope, that motion was a clue that the outer planets—those farthest from the Sun—were unique. Mercury, Venus, and Mars might seem inhospitable, forbidding, and downright deadly, but our sister terrestrial planets have more in common with Earth than these giants of the solar system's farthest reaches. The jovian planets are truly *other-worldly*, many times larger and more massive than Earth, yet less dense: They are balls of gas that coalesced around a dense core, accompanied by multiple moons and even rings.

In recent years, thanks to the *Hubble Space Telescope* and planetary probes such as the *Voyager* missions, *Galileo*, and the upcoming *Cassini* mission, the jovian planets and their moons have given up some of their mysteries. In this chapter, we explore these distant worlds. Their moons are so numerous and interesting that we will save a separate discussion of them for the chapter after this one.

The Jovian Line-Up

Jupiter was the supreme god in Roman mythology. His name was transmuted into Jove in Middle English, from which the word *jovial* comes, because, during the astrologically obsessed Middle Ages, the influence of Jupiter was regarded as the source of human happiness. Another adjective, *jovian*, usually spelled with a lowercase initial letter, means "like Jupiter." If the terrestrial planets have Earthlike ("terrestrial") qualities, all the jovian planets have much in common with Jupiter.

In addition to Jupiter, by far the largest planet in the solar system (about 300 times more massive than Earth and with a radius 11 times greater), the jovians include Saturn, Uranus, and Neptune.

Planetary Stats

The most immediately striking differences between the terrestrial and jovian worlds are in size and density. It is useful to recall our rough scale: If Earth is a golf ball 0.2 miles from the Sun, then Jupiter is a basketball 1 mile away from the Sun, and Pluto is a chickpea 8 miles away. At this scale, the Sun's diameter would be as big as the height of a typical ceiling (almost 10 feet). Although the jovian planets dwarf the terrestrials, they are much less dense. Let's sum up the jovians, compared to Earth:

Planet	Mass in Kilograms	Radius in Miles (and km)	"Surface" Magnetic Field (Relative to Earth)	"Surface" Gravity (Relative to Earth)	Rotation Period in Solar Days	Atmospheric Temperature in K
Earth	6.0×10^{24}	3,968 (6,380 km)	1.00	1.00	1.00	290
Jupiter	1.9×10^{27}	44,020 (71,400 km)	14	2.5	0.41	124
Saturn	5.7×10^{26}	37,200 (60,200 km)	0.7	1.1	0.43	97
Uranus	8.8×10^{25}	16,120 (25,600 km)	0.7	0.9	−0.72*	58
Neptune	1.0×10^{26}	15,500 (24,800 km)	0.4	1.2	0.67	59

*The rotation period of Uranus is negative because it is retrograde; like Venus, it rotates on its axis in the opposite direction from the other planets.

We have given a gravitational force and a temperature at the "surface" of the jovian planets, but as you'll see, they don't really have a surface in the sense that the terrestrial planets do. These numbers are the values for the outer radius of their swirling atmosphere. One surprise might be that the surface gravity of Saturn, Uranus, and Neptune is very close to what we have at the surface of Earth. Gravitational force depends on two factors, you'll recall: mass and radius. Although these outer planets are much more massive, their radii are so large that the force of gravity at their "surfaces" is close to that of the smaller, less massive Earth.

Jupiter's atmosphere is host to storms (left panel) and lightning (right panels) much like those found on Earth. These images from the Galileo *mission to Jupiter show a number of lightning strikes visible in the two images on the right, taken minutes apart.*

(Image from JPL/NASA)

Of the jovians, Jupiter and Saturn have the most in common with one another. Both are huge, with their bulk consisting mainly of hydrogen and helium. During the early phases of solar system development, the outer solar system (farther from the Sun) contained more water and organic materials than the inner part, and the huge mass and cooler temperatures of the outer planets meant that they were able gravitationally to hold on to the hydrogen and helium in their atmospheres.

Astronomer's Notebook

Why are the jovian planets so large? The jovian planets formed from nebular material (the stuff that formed the solar system) far from the Sun, in regions that were relatively cool. The lower temperatures enabled water, ice, and other molecules to condense in the outer solar system sooner, when those same molecules were still gaseous in the region closer to the still-forming protosun. As a result, there was a larger quarry of solid material available in the outer solar system to start building up into planetesimals. In effect, the outer solar system got a head start. These larger planetesimals (and their stronger gravitational fields) held on to a larger mass of hydrogen and helium, and thus the gas giants were born.

The terrestrials consist mostly of rocky and metallic materials, and the jovian planets primarily of lighter elements. The density of a planet is determined by dividing its mass by its volume. Although the outer planets are clearly much more massive (which, you might think, would make them more dense), they are much larger in radius, and so encompass a far greater volume. For that reason, the outer planets have (on average) a lower density than the inner planets, as is made clear in the following table:

Planet	Density (kg/m^3)
Earth	5,500
Jupiter	1,330
Saturn	710
Uranus	1,240
Neptune	1,670

Now, on to Uranus and Neptune—distant, faint, and unknown to ancient astronomers.

Although they are both much larger than Earth, they are less than half the diameter of Jupiter and Saturn; in our scale model, they would be about the size of a cantaloupe. Uranus and Neptune, though less massive than Saturn, are significantly more dense. Neptune is more dense than Jupiter as well, and Uranus approaches Jupiter in density. Take a look at the following "Astronomer's Notebook" sidebar to understand why this is so.

> ### Astronomer's Notebook
>
> To understand the difference between diameter and volume, think of our scale system of the planets. The difference in diameter between Earth and Jupiter tells you how many golf balls you can place next to the basketball to equal its diameter. The difference in *volume* between Earth and Jupiter tells you how many golf balls you could crush into dust and then fit *inside* the basketball. Volume is proportional to radius cubed (radius to the power of 3). So if object A has twice the radius of object B, object A has $2 \times 2 \times 2 = 8$ times B's volume!

Ponder Neptune. Remember, density is equal to the mass of an object divided by its volume. Although the mass of Neptune is about 19 times smaller than that of Jupiter, its volume is 24 times smaller. Thus, we expect its density to be about $^{24}/_{19}$ or 1.3 times greater. Although we cannot yet peer beneath the atmospheric surfaces of Uranus and Neptune, the higher densities of these two planets provide a valuable clue to what's inside.

Reflecting their genesis, all of the jovian planets have thick atmospheres of hydrogen and helium covering a super dense core slightly larger than Earth or Venus. The rocky cores of all four of the jovian planets are believed to have similar radii, on the order of 4,300 to 6,200 miles (7,000 to 10,000 km); but this core represents a much smaller fraction of the full radius of Jupiter and Saturn than do the cores of smaller Uranus and Neptune—thus the higher average density of the latter two planets.

The atmospheres of the jovian planets are ancient, and they've probably changed little since early in the lifetime of the solar system. With their strong gravitational fields and great mass, these planets have held on to their primordial atmospheric hydrogen and helium, whereas most of these elements long ago escaped from the less massive terrestrial planets, which have much weaker gravitational pull.

But here's where things get strange. On Earth, we have the sky (and atmosphere) above, and the solid ground below. In the case of the jovian planets, the gaseous atmosphere never really ends. It just becomes denser with depth, as layer upon layer presses down.

There is no "normal" solid surface to these planets! As the gases become more dense, they become liquid, which is presumably what lies at the core of the jovian worlds. When astronomers speak of the "rocky" cores of these planets, they are talking about chemical composition rather than physical state. Even on Earth, rock can be heated and pressed sufficiently to liquefy it (think of volcanic lava). Thus it is deep inside the jovian atmospheres: They become increasingly dense, but never solid, surrounding a liquid core. In the case of Jupiter and Saturn, the pressures are so great that even the element hydrogen takes on a liquid metallic form.

Latecomers: Uranus and Neptune

Since ancient times, the inventory of the solar system was clear and seemingly complete: the Sun and, in addition to Earth, five planets, Mercury, Venus, Mars, Jupiter, and Saturn. Then came along one of those scientific busybodies that the eighteenth century produced in abundance. Johann Daniel Titius, or Tietz (1729–1796), a Prussian born in what is now Poland, poked his curious nose into everything. He was a physicist, biologist, and astronomer who taught at the University of Wittenberg.

It occurred to him, in 1766, that the spacing of the planetary orbits from the Sun followed a fairly regular mathematical sequence. He doubled a sequence of numbers beginning with 0 and 3, like this: 0, 3, 6, 12, 24, 48, and so on.

He added 4 to each number in the sequence, and then divided each result by 10. Of the first seven answers Titius derived—0.4, 0.7, 1.0, 1.6, 2.8, 5.2, 10.0—six very closely approximated the relative distances from the Sun, expressed in astronomical units (remember, an A.U. is the mean distance between Earth and the Sun), of the six known planets.

No one paid much attention to Titius's mathematical curiosity until another Prussian astronomer, Johann Bode (1747–1826), popularized the sequence in 1772.

Neat as it was, the sequence, which became known as the Titius-Bode Law or simply Bode's Law, is now generally thought to be nothing more than numerology. For one thing, there is no planet at 2.8 A.U. (although this gap was filled later by the discovery of the asteroid belt at this location). Although the rule gives a number that is close to Uranus, it breaks down for the positions of Neptune and Pluto. Because those planets were as yet undiscovered in Bode's day, no one saw it as a problem, and astronomers began to think about the numbers beyond 10.0 A.U. They wondered if the Titius-Bode Law predicted other, as yet unknown, planets.

Astronomer's Notebook

Finding planets has gotten a lot easier since the days of Bode and Herschel. If you want to know the right ascension, declination, altitude, and azimuth of any of the planets as viewed from anywhere on the surface of Earth, point your Web browser to the planetary position calculator at www.imagiware.com/astro/planets.cgi.

The people of our planet didn't have to wait long to find out. On March 13, 1781, the great British astronomer William Herschel, tirelessly mapping the skies with his sister Caroline, took note of what he believed to be a comet in the region of a star called H Geminorum. On August 31 of the same year, a mathematician named Lexell pegged the orbit of this "comet" at 16 A.U.: precisely the next vacant slot the Titius-Bode Law had predicted.

Herschel, with the aid of a telescope, had discovered the first new planet—Uranus—since ancient times.

After the planet had been found, a number of astronomers began plotting its orbit. But something was wrong. Repeatedly, over the next half century, the planet's observed positions didn't totally coincide with its mathematically predicted positions. By the early nineteenth century, a number of astronomers began speculating that the new planet's apparent violation of Newton's laws of motion had to be caused by the influence of some undiscovered celestial body—that is, yet another planet. For the first time, Isaac Newton's work was used to identify the irregularity in a planet's orbit and to predict where another planet should be. All good scientific theories are able to make testable predictions, and here was a golden opportunity for Newton's theory of gravity.

Astro Byte

William Herschel, a court astronomer, knew what side his bread was buttered on and wanted to name the new planet Georgium Sidus—George's Star—after his patron, King George III of England. Bode intervened to suggest Uranus—in Roman mythology, the father of Saturn.

His sister, Caroline Herschel, was an accomplished astronomer in her own right. She discovered eight comets and three nebulae, and reduced and published her brother William's observations of nebulae after his death.

NASA's Hubble Space Telescope *made this image of Uranus, including its rings, the inner moons, and distinct clouds in Uranus's southern hemisphere.*

(Image courtesy of K. Seidelmann/USNO/NASA)

On July 3, 1841, John Couch Adams (1819–1892), a Cambridge University student, wrote in his diary: "Formed a design in the beginning of this week of investigating, as soon as possible after taking my degree, the irregularities in the motion of Uranus … in order to find out whether they may be attributed to the action of an undiscovered planet beyond it …." True to his word, in 1845, he sent to James Challis, director of the Cambridge Observatory, his calculations on where the new planet, as yet undiscovered, could be found. Challis passed the information to another astronomer, George Airy, who didn't get around to doing anything with the figures for a year. By that time, working with calculations supplied by another astronomer (a Frenchman named Jean Joseph Leverrier), Johann Galle, of the Berlin Observatory, found the planet that would be called Neptune. The date was September 23, 1846.

Earthbound Views: Uranus and Neptune

It is possible for the amateur astronomer to see both Uranus and Neptune. In fact, if you know where to look, Uranus is visible, albeit very faintly, even to the naked eye, provided that the night is very dark, very clear, and you are far from sources of light pollution. To view Neptune, which is much fainter than Uranus, requires an advanced amateur or professional telescope. You don't have to spend years sweeping the skies to find these dim and distant worlds. Consult one of the annual guides or websites listed in this chapter or in Appendix E for accurate information on positions of these planets.

But what can you expect to see? Uranus will appear as a greenish disk, probably featureless—though it is not impossible, given a very good telescope and superb atmospheric conditions, to see atmospheric features and bright spots. It is even possible to see Titania and Oberon, the largest of the planet's five moons.

Stormy weather on Neptune. Made with a combination of observations made with the Hubble Space Telescope (HST) *and the NASA Infrared Telescope Facility (ITF) on Mauna Kea, Hawaii, this image shows a surface mottled with storms and torn by winds of up to 900 miles per hour.*

(Image from L. Sromovsky/UW-Madison/NASA)

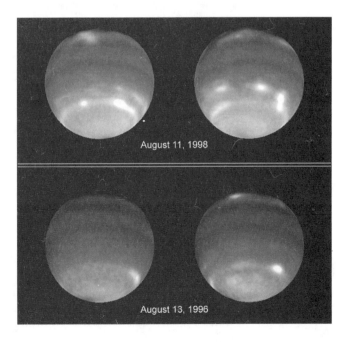

August 11, 1998

August 13, 1996

Even many advanced amateur astronomers have not seen Neptune. Blue in color, it is aptly named for the Roman god of the sea. If you locate the planet at all, it will be a featureless disk.

Earthbound Views: Jupiter and Saturn

In contrast to Uranus and Neptune, Jupiter and Saturn make for easy viewing. On a good, dark night, even a quite modest telescope will reveal the planets' belts (dark brownish stripes). The use of colored filters can enhance bands in Jupiter's atmosphere. Moreover, Jupiter rotates so fast (its day consumes a mere 10 hours) that any details you see will perceptibly move across the planet's face if you observe long enough. Its rapid rotation also makes the planet appear noticeably *oblate* (elongated), and it is even possible to observe the moons (such as Io) emerging from behind Jupiter as they orbit.

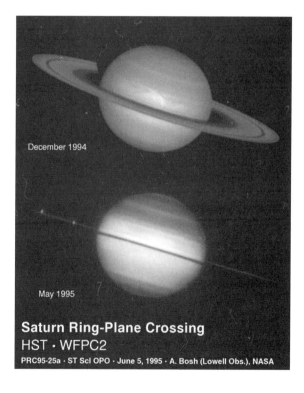

December 1994

May 1995

Saturn Ring-Plane Crossing
HST · WFPC2
PRC95-25a · ST ScI OPO · June 5, 1995 · A. Bosh (Lowell Obs.), NASA

Approximately every 15 years, Earth passes through Saturn's ring plane. The last passage (shown in the bottom image) occurred on May 22, 1995. The next crossing won't be until 2010.

(Image courtesy of NASA/STScI)

Although smaller and nearly twice as distant as Jupiter—and therefore appearing much smaller and dimmer than the larger planet—the sight of Saturn through a decent amateur telescope is thrilling. Expect to see the planetary disk and its belts (dark brownish stripes) and zones (brighter stripes), as well as its celebrated rings. You might even catch a glimpse of the moons, including Titan, the brightest and biggest of Saturn's nine moons (which we will discuss in the next chapter). Titan's atmospheric pressure is similar to Earth's, although its composition and temperature are different. Titan is slightly larger in diameter than the planet Mercury.

Close Encounter

The dark brownish stripes across Jupiter are called *belts,* the brighter stripes are called *zones.* Belts are dark, cooler regions, settling lower into the atmosphere as part of a convective cycle. Zones are regions of rising hot atmospheric gas. The bands are the result of regions of the atmosphere moving from high pressure to low pressure regions (much as they do on Earth). The rapid rotation of Jupiter confines this movement to narrow belts. The planet does have an atmospheric geography that can be mapped:

♦ The light-colored central band is the equatorial zone. It might appear white, orange, or yellow.

♦ The Great Red Spot, a hurricane that has been observed south of the equatorial zone since the invention of telescopes, has a diameter approximately twice that of Earth.

♦ On either side of the equatorial zone are dark bands called the north and south equatorial belts. At times, you might witness a south equatorial belt disturbance: an atmospheric storm.

♦ North and south of the equatorial belts are the *north* and *south temperate belts.*

♦ At the extreme northern and southern ends of Jupiter are the *polar regions,* which are sometimes barely perceptible and sometimes quite apparent dark areas.

Views from the *Voyagers* and *Galileo*

During the 1970s and 1980s, two *Voyager* space probes gave us unprecedented images of the jovian planets. *Voyager 1* visited Jupiter and Saturn, and *Voyager 2* added Uranus and Neptune to the list.

The *Voyager* missions also revealed volcanic activity on Io, one of Jupiter's moons. As for Saturn, a new, previously unknown system of rings emerged: several thousand ringlets. Ten additional moons were discovered orbiting Uranus, which also revealed the presence of a stronger magnetic field than had been predicted. And the Neptune flyby led to the discovery of three planetary rings as well as six previously unknown moons. The hitherto featureless blue face of the planet was resolved into atmospheric bands, as well as giant cloud streaks. As a result of the *Voyager 2* flyby, the magnetospheres of Neptune and Uranus were detected. As with the *Van Allen belts* around Earth, the magnetospheres of these planets trap charged particles (protons and electrons) from the solar wind.

If only its namesake could have lived to see it. Launched in 1989, *Galileo* reached Jupiter in 1995 and began a complex 23-month orbital tour of the planet and its moons (see Chapter 10) almost 400 years after the Italian astronomer first gazed on its colored bands and moons. Among the most extraordinary of *Galileo*'s discoveries is a new ring of dust that has a retrograde (backward) orbit around Jupiter. About 700,000 miles (1,120,000 km) in diameter, this doughnut-shaped ring moves in the opposite direction of the rotating planet and its moons. The *Galileo* mission ended dramatically on September 21, 2003, when the spacecraft entered Jupiter's atmosphere. It transmitted briefly before succumbing to high temperatures and pressures.

Star Words

The **Van Allen belts,** named for their discoverer, American physicist James A. Van Allen, are vast doughnut-shaped zones of highly energetic charged particles that are trapped in the magnetic field of Earth. The zones were discovered in 1958.

Close Encounter

Late in July 1994, *Galileo* was in orbit on the far side of Jupiter when more than 20 fragments of Comet Shoemaker-Levy 9 plunged into the atmosphere over a 6-day period. The fragments were the result of Jupiter's tidal forces that pulled a previously normal comet into a chain of smaller comets. Traveling at more than 40 miles per second (60 km/s), the fragments created a series of spectacular explosions in the planet's upper atmosphere, each with a force comparable to the detonation of a billion atomic bombs. So-called "black-eye" impact sites were created, the result of changes that the impacts produced in the upper atmosphere of the planet.

Rotation: A New Twist

With all the bands and surface features of the biggest jovian planets, you'd think it would be relatively easy to calculate rotation rates "by eye." Just look for a prominent surface feature and time how long it takes that feature to make one trip around.

Well, it's not so easy. Because these planets lack solid surfaces, different features on the surface actually rotate at differing rates! This *differential rotation* is not dramatic in the case of Jupiter, whose equatorial region rotates only slightly faster than regions at higher latitudes. East-west winds move at about 190 miles per hour (300 km/h) in Jupiter's equatorial regions, and at a zippy 800 miles per hour (1,300 km/h) in the equatorial regions of Saturn. It turns out that the best way to clock the rotation rates of these planets is not to look at their atmospheres, but to measure something tied to the planets' cores. The periods of fluctuation in the radio emission (which arise from the core-generated magnetic fields) are taken to be the "true" rotation rate.

Star Words

Differential rotation is a property of anything that rotates, and is not rigid. A spinning CD is a rigid rotator. A spinning piece of gelatin is not as rigid, and a spinning cloud of gas is even less so. The atmospheres of the outer planets and of the Sun have equatorial regions that rotate at a different rate from the polar regions.

Although Neptune and Saturn are slightly tipped on their axes similar to Earth (30, 27, and 23.5 degrees, respectively), Jupiter's axis is nearly perpendicular to the plane of its orbit; the planet tilts from the perpendicular a mere 3 degrees.

The true oddball in this respect is Uranus, which tilts 98 degrees, in effect lying on its side. The result of this peculiarity is that Uranus has the most extreme seasons in the solar system. While one pole experiences continuous daylight for 42 Earth years at a stretch, the other is plunged into an equal period of darkness.

Stormy Weather

Jupiter's spectacular surface features belie a turbulent atmosphere. In addition to a prevailing eastward and westward wind flow called *zonal flow*, there are many smaller-scale weather patterns, as evidenced by such features as the Great Red Spot.

The Great Red Spot

The Great Red Spot was first reported by the British scientist Robert Hooke (1635–1703). It is a storm, a swirling hurricane or whirlpool, of gigantic dimensions (twice the size of Earth), at least 300 years old. It rotates once every six days and is accompanied by other smaller storms. Neptune has a similar storm called the Great Dark Spot.

How could a storm last for three centuries or more? We know from our experience on Earth that hurricanes form over the ocean and might remain active there for days or weeks. When they move over land, however, they are soon spent (albeit often destructively); the land mass disrupts the flow pattern and removes the source of energy. On Jupiter, however, there is no land. Once a storm starts, it continues indefinitely, until a larger storm disrupts it. The Great Red Spot is currently the reigning storm on the planet.

Star Words

Zonal flow is the prevailing east-west wind pattern that is found on Jupiter.

Bands of Atmosphere

The atmospheric bands that are Jupiter's most striking feature are the result of *convective motion* and zonal wind patterns. Warm gases rise, while cooler gases sink. The location of particular bands appears to be associated with the wind speed on Jupiter at various latitudes.

Star Words _____

Convective motion is any flow pattern created by the rising movement of warm gases (or liquids) and the sinking movement of cooler gases (or liquids). Convective motion is known to occur in planetary atmospheres, the Sun's photosphere, and the tea kettle boiling water on your stove.

Anyone who watches an earthly television weather forecast is familiar with high-pressure and low-pressure areas. Air masses move from high-pressure regions to low-pressure regions. But we never see these regions on Earth as regular zones or bands that circle the planet. That's because Earth doesn't rotate nearly as fast as Jupiter. The rapid rotation of the gas giant spreads the regions of high and low pressure out over the entire planet.

Layers of Gas

On July 13, 1995, near the start of its mission, *Galileo* released an atmospheric probe, which plunged into Jupiter's atmosphere and transmitted data for almost an hour before it was destroyed by intense atmospheric heat and pressure. After analysis of this data (and earlier data from *Voyager*), astronomers concluded that Jupiter's atmosphere is arranged in distinct layers. Because there is no solid surface to call sea level, the *troposphere* (the region containing the clouds we see) is considered zero altitude, and the atmosphere is mapped in positive and negative distances from this.

Astro Byte _____

The total thickness of Saturn's three cloud layers—ammonia ice, ammonium hydrosulfide ice, and water ice—is about 155 miles (250 km), compared to about 50 miles (80 km) on Jupiter.

Just above the troposphere is a haze layer, and just below it are white clouds of ammonia ice. Temperatures in this region are 125–150 K. Starting at about 40 miles (60 km) below the ammonia ice level is a cloud layer of ammonium hydrosulfide ice, in which temperatures climb to 200 K. Below this level are clouds of water ice and water vapor, down to about 60 miles (100 km). Farther down are the substances that make up the interior of the planet: hydrogen, helium, methane, ammonia, and water, with temperatures steadily rising the deeper we go. Molecules at these depths can be probed with observations at radio wavelengths.

Saturnine Atmosphere

Saturn's atmosphere is similar to Jupiter's—mostly hydrogen (92.4 percent) and helium (7.4 percent) with traces of methane and ammonia—however, its weaker gravity results in thicker cloud layers that give the planet a more uniform appearance, with much subtler banding, than Jupiter. Temperature rises much more slowly as a function of depth in the atmosphere on Saturn.

The Atmospheres of Uranus and Neptune

The atmospheres of Uranus and Neptune have not been probed by unmanned space vehicles, but they have been studied spectroscopically from Earth, revealing that, like Jupiter and Saturn, they are mostly hydrogen (about 84 percent) and helium (about 14 percent). Methane makes up about 3 percent of Neptune's atmosphere and 2 percent of Uranus's, but ammonia is far less abundant on either planet than on Jupiter and Saturn. Because Uranus and Neptune are colder and have much lower atmospheric pressure than the larger planets, any ammonia present is frozen. The lack of ammonia in the atmosphere and the significant presence of methane give both Uranus and Neptune a bluish appearance, because methane absorbs red light and reflects blue. Uranus, with slightly less methane than Neptune, is blue-green, while Neptune is quite blue.

Uranus reveals almost no atmospheric features. The few present are submerged under layers of haze. Neptune, as seen by *Voyager 2*, reveals more atmospheric features and even some storm systems, including a Great Dark Spot, an area of storm comparable in size to Earth. Discovered by *Voyager 2* in 1989, the Great Dark Spot had vanished by the time the *Hubble Space Telescope* observed the planet in 1994.

Inside the Jovians

How do you gather information about the interior of planets that lack a solid surface and that are so different from Earth? You combine the best observational data you have with testable, constrained speculation known as theoretical modeling. Doing just this, astronomers have concluded that the interiors of all four jovians consist largely of the elements found in their atmospheres: hydrogen and helium. Deeper in the planets, the gases, at increasing pressure and temperature, become liquid, but never solid.

In the case of Jupiter, it is believed that the hot liquid hydrogen is transformed from molecular hydrogen to metallic hydrogen and behaves much like a molten metal, in which electrons are not bound to a single nucleus, but move freely, conducting electrical charge. As you shall see in just a moment, this state of hydrogen is likely related to the creation of Jupiter's powerful magnetosphere—the result of its strong magnetic field. Saturn's internal composition is doubtless similar to Jupiter's, although its layer of metallic hydrogen is probably proportionately thinner, and its core is slightly larger. Temperature and pressure at the Saturnine core are certainly less extreme than those on Jupiter.

Uranus and Neptune are believed to have rocky cores of similar size to those of Jupiter and Saturn surrounded by a slushy layer consisting of water clouds and, perhaps, the ammonia that is largely absent from the outer atmosphere of these planets. Because Uranus and Neptune have significant magnetospheres, some scientists speculate that the ammonia might create an electrically conducting layer, needed to generate the detected magnetic field. Above the slushy layer is molecular hydrogen. Without the enormous internal pressures present in Jupiter and Saturn, the hydrogen does not assume a metallic form.

The Jovian Magnetospheres

Jupiter's magnetosphere is the most powerful in the solar system. Its extent reaches some 18,600,000 miles (30 million km) north to south. Saturn has a magnetosphere that extends about 600,000 miles (1 million km) toward the Sun. The magnetospheres of Uranus and Neptune are smaller, weaker, and (strangely) offset from the gravitational center of the planets.

The rapid rate of rotation and the theorized presence of electrically conductive metallic hydrogen inside Jupiter and Saturn account for the generation of these planets' strong magnetic fields. Although Uranus and Neptune also rotate rapidly, it is less clear what internal material generates the magnetic fields surrounding these planets, because they are not thought to have metallic hydrogen in their cores. With charged particles trapped by their magnetospheres, the jovian planets experience *Aurora Borealis*, or "Northern Lights," just as we do here on Earth. These "lights" occur when charged particles escape the magnetosphere and spiral along the field lines onto the planet's poles. The *Hubble Space Telescope* has imaged such auroras at the poles of Jupiter and Saturn.

Jupiter as observed with the Very Large Array (VLA) in Socorro, New Mexico. This image of the 20-cm radiation from the planet highlights the powerful magnetic field of the solar system's largest planet.

(Image courtesy of Imke de Pater)

The Least You Need to Know

♦ The jovian planets are Jupiter, Saturn, Uranus, and Neptune.

♦ Although they are the largest, most massive planets in the solar system, the jovian planets are on average less dense than the terrestrial planets. Their outer layers of hydrogen and helium gas cover a dense core.

♦ Uranus was discovered in 1781, partly because of a bit of numerology called the Bode (or Titius-Bode) Law.

♦ Neptune was discovered in 1846, because astronomers were searching for an explanation of Uranus's slightly irregular orbit. Newton's theory of gravity provided an explanation—the mass of another planet.

♦ The jovian planets all have in common thick atmospheres, ring systems, and strong magnetic fields.

♦ The missions to the outer planets, *Voyager 1*, *Voyager 2*, and *Galileo*, have had a huge impact on increasing our understanding of the jovian planets. The *Cassini* mission to Saturn and its moons, which will arrive in 2004, is likely to have a similar effect.

Chapter **10**

The Far End of the Block

In This Chapter

- ◆ Anatomy of Saturn's rings
- ◆ The other jovian ring systems
- ◆ The large moons of Jupiter and the other jovians
- ◆ Europa and Titan: Is life possible there?
- ◆ Atmospheres of the jovian moons
- ◆ Tiny and distant: Pluto and Charon

Among the great showpieces of our solar system are the fabulous rings of Saturn. Visible even through modest binoculars, Saturn's ring system appears as solid as the planet. The truth is that neither the planet nor the rings are solid. The planet is a "gas giant," and the rings are composed of small bits of ice and rock. Perhaps more surprisingly, the *Voyager* spacecraft and ground-based adaptive optics images show that all of the outer planets have ring systems—not just Saturn.

If you were somehow disappointed to discover, in the preceding chapter, that the jovian planets are just big balls of gas, you can take satisfaction in knowing that some of their moons at least are solid, large, and much more like the terrestrial planets. Jupiter's Ganymede and Saturn's Titan, for example, are larger than the planet Mercury. And two of the moons—Jupiter's Europa and Saturn's Titan—present some of the most promising locations for life, outside of Earth, in the solar system.

And then there is Pluto. Unknown before 1930 because of its great distance and diminutive size (it is smaller than Earth's moon and similar in size to many of the jovian moons), Pluto is unlike either the terrestrial or jovian worlds. But we first have unfinished business from the preceding chapter. Let's begin with the rings of the outer planets.

Lord of the Rings

The spectacular rings of Saturn, a thrill to view in remarkable images from *Voyager 2* and the *Hubble Space Telescope*, are even more exciting when seen through your own telescope.

Looking from Earth

Galileo's telescope, a wondrous device in the early seventeenth century, would be no match for an inexpensive amateur instrument today. When he first observed the planet, all Galileo could tell about Saturn was that it seemed to have "ears." He speculated that this feature might be topographical, great mountain ranges of some sort. Or, he thought, perhaps Saturn was a triple planet system, with the "ears" as outrigger planets. It wasn't until a half-century after Galileo's speculations, in 1656, that Christian Huygens, of the Netherlands, was able to make out this feature for what it was: a thin ring encircling the planet. A few years later, in the 1670s, the Italian-born French astronomer Gian Domenico Cassini (1625–1712) discovered the dark gap between what are now called rings A and B. This gap is called the *Cassini division*.

Star Words

The **Cassini division** is a dark gap between rings A and B of Saturn. It is named for its discoverer, Gian Domenico Cassini (1625–1712). The NASA mission named in his honor is slated to arrive at Saturn in 2004.

In all, six major rings, each lying in the equatorial plane of Saturn, have been identified, of which three, in addition to the Cassini division and a subtler demarcation called the Encke division, can be seen from Earth with a good telescope. With a typical amateur instrument you should be able at least to see ring A (the outermost ring), the Cassini division, and inside the Cassini division, ring B.

If you become a serious Saturn observer, you will notice that the rings of Saturn are seen at different angles at different times. Most of the time we look down at an angle on the ring system, and rarely we see it "edge-on" (an image of Saturn seen "edge-on" is included in the previous chapter). When the angle is right, it is possible to see the dramatic image of Saturn's shadow cast onto its rings. Consult any of the guides in Appendix E for information on where to look for Saturn and when to view it.

The rings most readily visible from Earth are vast, the outer radius of the A ring stretching more than 84,800 miles.

Big as they are, the rings are also very thin—in places only about 65 feet (20 m) thick. If you wanted to make an accurate scale model of the rings and fashioned them to the thickness of this sheet of paper, they would have to be a mile wide to maintain proper scale.

Astronomer's Notebook

The point at which a moon orbiting a planet might be pulled apart is not random. Edouard Roche, a nineteenth-century mathematician, calculated the distance as 2.4 times the radius of the planet, assuming the moon is of density similar to that of the parent planet. Saturn's "Roche limit" is 89,280 miles (144,000 km) from the planet's center, 2.4 times its radius. Any moon that crosses this boundary will be pulled to pieces.

Speculation as to the composition of the rings began with their discovery in the mid-seventeenth century. In 1857, the British physicist James Clerk Maxwell speculated that the rings must consist of many small particles in orbit around Saturn. By the end of the century, the instrumentation existed to measure reflectivity, the differences in the way sunlight was reflected from the rings. These observations showed that the rings behaved as was to be expected if they were made up of particles; that is, orbital speeds closer to the planet were faster than those farther out. In other words, the rings were in differential rotation, not rotating as a solid disk might.

Where do the rings come from?

There are two ways to think about the question, and both involve the gravitational field of the host planet. First, the rings might be the result of a shattered moon. According to this theory, a satellite could have been orbiting too close to the planet and been torn apart by tidal forces (the same sort of forces that pulled comet Shoemaker-Levy 9 to pieces), or the moon might have been shattered by a collision. In either case, the pieces of the former moon continued to orbit the planet, but now as fragmentary material.

Saturn and its rings as seen with the Very Large Array at 3.6 cm. The cooler ring system is seen in absorption (light) against the dark emission from the upper atmosphere of the planet.

(Image credit B. Dunn/NRAO/ AUI)

The other possibility is that the rings are material left over from the formation of the planet itself, material that was never able to coalesce into a moon due to the strong gravitational field of the host planet.

Astronomer's Notebook

The biggest gap in Saturn's rings, the Cassini division, was discovered long before the *Voyager* missions. It is the result not of moonlets, but of the gravitational influence of Mimas, Saturn's innermost moon, which deflected some of the ring's particles into different orbits, creating a gap large enough to be visible from Earth.

Looking with *Voyager*

Zooming in close, the *Voyager* probes told us much more about the rings than we could have discovered from our earthly perspective. First, data from *Voyager* confirmed that the rings are indeed made up of particles, primarily of water ice. *Voyager* also revealed additional rings, invisible from Earth. The F ring, for example, is more than twice the size of the A ring, stretching out to 186,000 miles.

The D ring is the innermost ring—closer to the planet than the innermost ring visible from Earth, the C ring. F and E are located outside the A ring.

But these additional rings are only part of what *Voyager* told us. *Voyager 2* revealed that the six major rings are composed of many thousands of individual ringlets, which astronomers liken to ripples or waves in the rings.

Voyager 2 also revealed many gaps within the rings, which are believed to be caused by small moonlets, which might be considered very large ring particles—a few miles in diameter—in orbit around Saturn. The gaps are, in some sense, the wakes of these bodies.

Perhaps the most remarkable *Voyager* discovery concerning Saturn's rings is regarding the outermost F ring. Its structure is highly complex, sometimes appearing braided. Apparently, the structure of the F ring is influenced by two small outlying moons that bracket the ring; these are called shepherd satellites, and they keep the F-ring particles from moving in or out.

More Rings on the Far Planets

During a 1977 Earth-based observation of Uranus in the course of a *stellar occultation* (the passage of Uranus in front of a star), the star's light dimmed several times before disappearing behind the planet. That dimming of the star's light revealed the presence of nine thin, faint rings around the planet. *Voyager 2* revealed another pair. Uranus's rings are very narrow—most of them less than 6 miles (10 km) wide—and are kept together by the kind of shepherd satellites that are found outside of Saturn's F ring. Neptune has faint rings similar to those of Uranus.

Star Words

A **stellar occultation** occurs when a planet or moon passes in front of a star, dimming the star's light (as seen from Earth). The exact way in which the light dims can reveal details, for example, in the occulting planet's or moon's atmosphere.

On the Shoulders of Giants

One of the key differences between the terrestrial and many jovian planets is that, although the terrestrials have few if any moons, the jovians have many: 28 (at least) for Jupiter, more than 28 for Saturn, 21 for Uranus, and 8 for Neptune. Of these known moons, only six are classified as large bodies, comparable in size to Earth's moon. Our own moon is all the more remarkable when compared to the moons of the much larger jovian planets. It is larger than all of the known moons except for Ganymede, Titan, Callisto, and Io. The largest jovian moons (in order of decreasing radius) are …

- **Ganymede,** orbiting Jupiter; approximate radius: 1,630 miles (2,630 km).

- **Titan,** orbiting Saturn; approximate radius: 1,600 miles (2,580 km).

- **Callisto,** orbiting Jupiter; approximate radius: 1,488 miles (2,400 km).

- **Io,** orbiting Jupiter; approximate radius: 1,130 miles (1,820 km).

- **Europa,** orbiting Jupiter; approximate radius: 973 miles (1,570 km).

- **Triton,** orbiting Neptune; approximate radius: 856 miles (1,380 km).

It is interesting to compare these to Earth's moon, with a radius of about 1,079 miles (1,740 km), and the planet Pluto, smaller than them all, with a radius of 713 miles (1,150 km).

The rest of the moons are either *medium-sized bodies*—with radii from 124 miles (200 km) to 465 miles (750 km)—or *small bodies*, with radii of less than 93 miles (150 km). In recent years, many small bodies have been discovered orbiting Jupiter and Saturn. Many of the moons are entirely or mostly composed of water ice, and some of the smallest bodies are no more than irregularly shaped rock and ice chunks.

Faraway Moons

Thanks to the *Voyager* and *Galileo* space probes, we have some remarkable images and data about the moons at the far reaches of the solar system. Those that have received the most attention, because they are the largest, are the so-called Galilean moons of Jupiter, Saturn's Titan, and Neptune's Triton.

Jupiter's Four Galilean Moons

The four large moons of Jupiter are very large, ranging in size from Europa, only a bit smaller than Earth's moon, to Ganymede, which is larger than the planet Mercury. Certainly, they are large enough to have been discovered even through the crude telescope of Galileo Galilei, after whom they have been given their group name. In his notebooks, Galileo called the moons simply I, II, III, and IV. Fortunately, they were eventually given more poetic names, Io, Europa, Ganymede, and Callisto, drawn from Roman mythology. These four are, appropriately, the attendants serving the god Jupiter.

Io is closest to Jupiter, orbiting at an average distance of 261,640 miles (421,240 km); Europa comes next (416,020 miles or 669,792 km); then Ganymede (663,400 miles or 1,068,074 km); and finally Callisto (1,165,600 miles or 1,876,616 km). Intriguingly, data from *Galileo* suggests that the core of Io is metallic, and its outer layers rocky—much like the planets closest to the Sun. Europa has a rocky core, with a covering of ice and water. The two outer large moons, Ganymede and Callisto, also have icy surfaces surrounding rocky cores.

This pattern of decreasing density with distance from the central body mimics that of the solar system at large, in which the densest planets, those with metallic cores, orbit nearest the Sun, while those composed of less dense materials orbit farthest away. This similarity is no mere coincidence and can be used to investigate how the Jupiter "system" formed and evolved.

Let's look briefly at each of Jupiter's large moons.

Because of our own moon, we are accustomed to thinking of moons generally as geologically dead places. Nothing could be further from the truth in the case of Io, which has the distinction of being the most geologically active object in the entire solar system.

A spectacular portrait of Jupiter's volcanic moon Io by the Galileo *spacecraft. The image was created on March 29, 1998, at a distance of 183,000 miles (294,000 kilometers) from Io.*

(Image from JPL/NASA)

Io's spectacularly active volcanoes continually spew lava, which keeps the surface of Io relatively smooth—any craters are quickly filled in—but also angry-looking, vivid orange and yellow, and sulfurous. In truth, Io is much too small to generate the kind of heat energy that produces volcanism (volcanic activity); however, orbiting as close as it does to Jupiter, it is subjected to the giant planet's tremendous gravitational field, which produces tidal forces (see Chapter 6). These forces stretch the planet from its spherical shape and create the geologically unsettled conditions on Io. Think about what happens when you rapidly squeeze a small rubber ball. The action soon makes the ball quite warm. The forces exerted on Io by Jupiter are analogous to this, but on a titanic scale. By the way, don't invest in an Io globe for your desk. Its surface features change faster than political boundaries on Earth!

In contrast to Io, Europa is a cold world—but probably not an entirely frozen one, and perhaps, therefore, not a dead world. Images from *Galileo* suggest that Europa is covered by a crust of water ice, which is networked with cracks and ridges. It is possible that beneath this frozen crust is an ocean of *liquid* water (not frozen water or water vapor). Liquid water is certainly a requisite of life on Earth, though the presence of water does not necessarily imply the existence of life. Still, the prospects are most exciting. Europa might be, in effect, a lifeboat in the outer solar system, although before we get our hopes up, we need to realize just how cold the surface of Europa is (130 K) and how thin its atmosphere is—at a pressure approximately one billionth that on Earth. Any life on Europa would have to be lurking deep in oceans that could be beneath its icy sheath.

Ice on the surface of Europa. The smooth dark regions might be areas where water has welled up from underneath the "ice shelf" that covers the moon.

(Image from NASA/JPL)

Ganymede is the largest moon in the solar system (bigger than the planet Mercury). Its surface shows evidence of subsurface ice that was liquefied by the impact of asteroids and then refrozen. Callisto is smaller but similar in composition. Both are ancient worlds of water ice, impacted by craters. There is little evidence of the current presence of liquid water on these moons.

Frozen waves? An area within the Valhalla region on Jupiter's moon, Callisto, with the Sun illuminating the surface from the left. Callisto's surface consists of rock and water ice.

(Image from NASA/JPL)

Titan: Saturn's Highly Atmospheric Moon

If Io is the most geologically active moon in the solar system and Ganymede the largest, Saturn's Titan enjoys the distinction of having the most substantial atmosphere of any moon. No wispy, trace covering, Titan's atmosphere is mostly nitrogen (90 percent) and argon (nearly 10 percent) with traces of methane and other gases in an atmosphere thicker than Earth's. Earth's atmosphere consists of 78 percent nitrogen, 21 percent oxygen, and 1 percent argon. Surface pressure on Titan is about 1.5 times that of Earth. But its surface is very cold, about 90 K. Remember 90 K is –183 C!

Titan's atmosphere prevents any visible-light view of the surface, although astronomers speculate that the interior of Titan is probably a rocky core surrounded by ice, much like Ganymede and Callisto. Because Titan's temperature is lower than that of Jupiter's large moons, it has retained its atmosphere. The presence of an atmosphere thick with organic molecules (carbon monoxide, nitrogen compounds, and various hydrocarbons have been detected in the upper atmosphere) has led to speculation that Titan could possibly support some form of exotic life. The *Cassini-Huygens* spacecraft will release a probe into the moon's atmosphere when it arrives at Saturn in 2004.

This artist's conception shows the Titan Probe *of the Cassini-Huygens Mission to the planet Saturn arriving on the surface of Titan.*

(Image from NASA)

> **Close Encounter**
>
> Launched from Kennedy Space Center on October 15, 1997, the *Cassini-Huygens* spacecraft consists of the *Cassini* orbiter, which will orbit Saturn and its moons for four years, and the *Huygens* probe, which will enter the atmosphere of Saturn's moon Titan and land on its surface. Seventeen nations contributed to building these spacecraft.
>
> *Cassini*'s orbital tour is expected to provide an unprecedented wealth of images and other data, giving scientists their fullest understanding of any jovian world. The *Huygens* mission will yield images and other important data as the probe directly samples Titan's atmosphere. You can check on the progress of the mission and view its photos of Jupiter at www.jpl.nasa.gov/cassini/.

Triton, Neptune's Large Moon

Triton's distinction among the jovian moons is a retrograde (backward) orbit—in the reverse direction of the other moons. Moreover, Triton is inclined on its axis about 20 degrees and is the only large jovian moon that doesn't orbit in the equatorial plane of its planet. Many astronomers believe that these peculiarities are the result of some violent event, perhaps a collision. Others suggest that Triton didn't form as part of the Neptunian system of moons, but was captured later by the planet's gravitational field.

Triton's atmosphere is so thin that *Voyager 2* had no trouble imaging the moon's surface, finding vast lakes of water ice or water-ammonia mixtures there. Nitrogen frost, found at the polar caps, appears to retreat and reform seasonally.

A Dozen More Moons in the Outer Solar System

Thanks to *Voyager*, the six medium-sized moons of Saturn have also been explored. All of these bodies are *tidally locked* with Saturn, their orbits synchronous, so that they show but one face to their parent planet. They are frozen worlds, made of mostly rock and water ice. The most distant from Saturn, Iapetus, orbits some 2,207,200 miles (3,560,000 km) from its parent. Because these moons orbit synchronously, astronomers speak of their *leading faces* and *trailing faces*. That one face always looks in the direction of the orbit and the other in the opposite direction has created asymmetrical surface features on some of these moons. The leading face of Iapetus, for example, is very dark in color, while the trailing face is quite light. Although some astronomers suggest that the dark material covering this moon's leading face is generated internally, others believe that Iapetus sweeps up the material it encounters in the course of its orbit.

The innermost moon of Saturn, Mimas is 115,320 miles (186,000 km) from Saturn. It is also the smallest of Saturn's moons, with a radius of just 124 miles (200 km). Mimas is very close to Saturn's rings and seems to have been battered by material associated with them. Heavily cratered overall, this small moon has one enormous crater named for the astronomer William Herschel, which makes it resemble the Empire's "Death Star" in *Star Wars*. Whatever caused this impact probably came close to shattering Mimas. Indeed, some astronomers believe that similar impacts might have created some of the debris that formed Saturn's great rings.

Star Words

Moons that are **tidally locked** to their parent planet have a **leading face** and a **trailing face**; the leading face always looks in the direction of its orbital motion, the trailing face away from it.

The medium-sized moons of Uranus are Miranda, orbiting 80,600 miles (130,000 km) above the planet; Ariel, 118,400 miles (191,000 km) out; Umbriel, 164,900 miles (266,000 km) out; Titania, 270,300 miles (436,000 km) out; and Oberon, 361,500 miles (583,000 km) out. Of these, the most remarkable is Miranda, which, in contrast to the other moons, is extremely varied geographically, with ridges, valleys, and oval-shaped faults. To the camera of *Voyager 2*, it presented a chaotic, violently fractured, cobbled-together surface unlike that of any other moon in the solar system. Clearly, this moon had a violent past, although it is unclear whether the disruptions it suffered came from within, without, or both. Some astronomers believe that Miranda was virtually shattered, its pieces coming back together in a near-jumble.

The Frankensteinian face of Miranda, one of Uranus's moons.

(Image from NASA)

> **Astro Byte** _____
>
> Pluto (called Hades in Greek mythology) is the Roman god of the dead, whose realm is eternal darkness. The name was also appealing for the newly discovered planet because its first two letters could be worked into an astrological symbol that combined P and L, the initials of the planet's "godfather," Percival Lowell.

Pluto Found

We are accustomed to thinking of the planets as our ancient companions, so it might come as a surprise to many that one planet, Pluto, was not found until 1930. In Chapter 9 we saw that irregularities in the orbit of Uranus implied the existence of Neptune, which wasn't discovered until the middle of the nineteenth century. Yet the discovery of Neptune never did fully account for the idiosyncrasies of the Uranian orbit. Neptune also seemed to be influenced by some as-yet-unknown body.

The keen, if eccentric, astronomer Percival Lowell (1855–1916) crunched the numbers for Uranus's orbit and searched for a new planet, in vain, for some 10 years. Clyde Tombaugh (1906–1997) was too young to have known Percival Lowell personally, but he took a job as assistant astronomer at the observatory Lowell had built in Flagstaff, Arizona. It was Tombaugh who eventually found the planet by studying photographic images in 1930.

That Pluto was found a mere six degrees from where Lowell had said it would be is more a testament to serendipity than to astronomical calculations. The supposedly persistent irregularities in the orbits of Uranus and Neptune, on which he based his calculations, don't exist.

> **Astro Byte** _____
>
> Pluto was inside Neptune's orbit from 1979 until 1999. The high inclination of Pluto's orbit (with respect to the plane of the solar system) means that a chance collision between these two planets is negligible.

But there was Pluto nevertheless: 3.7 billion miles (5.9 billion km) from the Sun, on average, but with an orbit so eccentric that it actually comes closer to the Sun than Neptune does about every 248 years. It has yet to be visited by unmanned probes (although the *Hubble Space Telescope* has taken pictures) and, with amateur telescope equipment, is next to impossible to detect. It is very faint and, therefore, virtually indistinguishable from surrounding stars, except for its movement.

Astro Byte

Astronomical objects (however minor) are typically named by their discoverers. In 1919, the International Astronomical Union (IAU) was given the task of bringing order to the then rather unorganized naming of surface features on Mars and the Moon. Since then, this organization has been the only internationally recognized authority on the naming of features on solar system objects and objects outside the solar system. For more on this topic, check out www.iau.org/IAU/Activities/nomenclature/.

A "New" Moon

If, having been discovered in 1930, Pluto was a late addition to our known solar system, its moon, Charon, is almost brand new, having been found in 1978. Named, fittingly, for the mythological ferryman who rowed the dead across the River Styx to the under-world ruled by Pluto, Charon is a little more than half the size of its parent: 806 miles (1,300 km) in diameter versus Pluto's 1,426 miles (2,300 km). Orbiting 12,214 miles (19,700 km) from Pluto, it takes 6.4 Earth days to make one circuit. Pluto and Charon are tidally locked—forever facing one another—the orbital period and rotation period for both are synchronized at 6.4 days. Like Venus and Uranus, Pluto's rotation is retrograde—spinning on its axis in the opposite direction of most of the other planets.

Pluto and its moon Charon as imaged with the Hubble Space Telescope *(HST) in 1994. The image was taken when Pluto was 2.6 billion miles (4.4 billion km) from Earth—nearly 30 A.U. away.*

(Image from R. Albrecht/ESA/NASA)

Where Did Pluto Come From?

Except that it doesn't orbit another planet—and, indeed, has a moon of its own—Pluto looks more like a jovian moon than a planet. It fits into neither the terrestrial nor jovian mold. Some astronomers think that Pluto might be a renegade moon, escaped from Neptune's gravitational influence due to a collision or interaction involving Triton, Pluto, Charon, and Nereid. Others regard it as a kind of spare part, something left over from the creation of the solar system, and perhaps only the largest of a number of such icy objects in the outer reaches of the solar system, the Kuiper Belt.

The Least You Need to Know

- The jovian realm is rich in moons and planetary rings. Some jovian moons are volcanically active, others have measurable atmospheres.

- Rings are characteristic of all the jovian planets, but only those of Saturn are readily seen from Earth without special optics.

- Planetary rings consist of small orbiting particles, perhaps fragments from pulverized moons that strayed too close to the parent planet and were pulled apart.

- Io has frequent volcanic eruptions that loft material into orbit around Jupiter. Europa, also orbiting Jupiter, might have liquid water beneath its frozen, cracked surface.

- Pluto, discovered in 1930, doesn't fit neatly into either the terrestrial or jovian category and is in many ways more like a jovian moon than a planet. It might be an escaped moon of Neptune or simply the largest object in the Kuiper Belt.

11

Worlds Near and Far

In This Chapter

- ◆ A theory of solar system origin and evolution
- ◆ The death of the solar system
- ◆ The search for other planetary systems
- ◆ Current and future techniques in the search

In this chapter, we explore planetary systems in a universal sense. It has become increasingly clear that our solar system is just one among myriad planetary systems. Only a decade ago, our own planetary system was the only one that we knew for sure existed. But times have changed. Of course, the closest planetary system is still our own. But observations since 1995 have revealed that we are not alone, at least not as a collection of planets. Planets appear to be commonplace around all sorts of stars.

One of the most difficult things about understanding the origins of our own solar system—the Sun and the collection of planets that orbit it—is that it has been around for a very long time. Based on studies of meteorites, astronomers believe the solar system is 4.6 billion years old. The problem is akin to looking at a middle-aged man and being asked to describe the moment of his birth. We might be able to watch other births happen, and assume that his birth was much the same. But no one now living was present to watch the birth of this solar system, so uncovering its beginnings has required some serious sleuthing.

If we consider observations from our own solar system as well as those of the over 100 other planetary systems that we are now aware of—many of them at very different states of evolution—we can better understand how planetary systems, including our own, are born, live, and die.

Solar System History

Even though no one was around to record the series of events that created the solar system, there are a few fragments that survive from those early moments that give us clues to how the planets took shape around the youthful Sun. The most important clues to the origin of the solar system are to be found not in the Sun and planets, but in those untouched smaller fragments: the asteroids, meteoroids, and some of the planetary moons (including our own).

What Do We Really Know About the Solar System?

In a very real sense we have—in meteorites and moon rocks—"witnesses" to the creation of the solar system. These geological remnants are relatively unchanged from the time the solar system was born. But how do we make up for an absence of precedents from which to draw potentially illuminating analogies?

One approach is to find another planetary system forming around a star younger than the Sun and draw analogies from it. The *Hubble Space Telescope*, among other spaceborne probes, has given us tantalizing clues about the formation of planetary systems. Around the star Beta Pictoris, for example, astronomers have imaged a disk of dusty material larger than the orbit of the most distant planet in our solar system, Pluto.

Beta Pictoris is believed to be a dusty disk surrounding a young star. The presence of this disk, imaged by the Hubble Space Telescope, *supports the nebular hypothesis. The light from the central star has been blocked so that the faint disk could be imaged.*

(Image from Al Schultz, CSC/STScI, and NASA)

The best way to understand the formation of our own solar system is to study both our own system, which is nearby, and the wide variety of other protoplanetary and planetary systems being discovered by astronomers using the world's best telescopes.

Let's start with what we know about the closest planetary system, our own. There are a number of fundamental things that we know about our solar system, and any explanation that we come up with must, at the very least, account for what we observe.

Here are some undeniable facts that the last 300 years of planetary exploration have given us:

◆ Most of the planets in the solar system rotate on their axis in the same direction as they orbit the Sun (counterclockwise as seen from the North Pole of Earth), and their moons orbit around them in the same direction.

◆ The planets in the inner reaches of the solar system are rocky and bunched together, and those in the outer part are gaseous and widely spaced.

◆ Most of the planets (with the exception of Pluto) orbit the Sun in elliptical paths that are very nearly circles.

◆ Except for the innermost planet (Mercury) and the outermost (Pluto), the planets orbit in approximately the same plane (near the ecliptic), and they all orbit in the same direction.

◆ Asteroids and comets are very old and are located in particular places in the solar system. Comets are found in the Kuiper Belt and Oort Cloud, and asteroids are in the asteroid belt between Mars and Jupiter.

In addition, it is clear that the asteroids that have been examined are some of the oldest unchanged objects in the solar system, and that comets travel in highly elliptical orbits, originating in the far reaches of the solar system. The most important conclusion we can draw from these observations is that the solar system appears to be fundamentally orderly rather than random. It doesn't appear that the Sun formed first, and then gradually captured its nine planets from surrounding space.

A Plausible Scenario

Here is a possible portrait of the formation of our solar system: A cloud of interstellar dust, measuring about a light-year across, begins to contract, rotating more rapidly the more it contracts. With the accelerating rotation comes a flattening of the cloud into a pancakelike disk, perhaps 100 A.U. across—100 times the current distance between Earth and the Sun.

The original gases and dust grains that had formed the nebular cloud have contracted into condensation nuclei, which begin to attract additional matter, forming clumps that rotate within the disk.

The clumps encounter other clumps and more matter, growing larger by accretion. Accretion is the gradual accumulation of mass, and usually refers to the building up of larger masses from smaller ones through the mutual gravitational attraction of matter.

Close Encounter

Startling new infrared and radio telescope evidence supporting the condensation theory was reported in April 1998 by astronomers working at the Keck Observatory in Mauna Kea, Hawaii, and at Cerro Tololo Inter-American Observatory in Chile. Studying a star known as HR4796A, 220 light-years from Earth, the astronomers discovered a vast dust disk forming around it. A doughnutlike hole, slightly larger than the distance between the Sun and Pluto, surrounds the star, and the disk itself extends more than twice the distance of the doughnut hole.

Although astronomers did not detect any planets in this very distant object, they believe the hole in the disk might have been caused by the gravitational force of one or more inner planets. In effect, then, astronomers believe they are seeing a distant planetary system in the making.

In 1984, a disk was observed around another star, Beta Pictoris; but astronomers are even more excited about the 1998 discovery because they are more certain of the young age of HR4796A than they are of the age of Beta Pictoris. At a mere 10 million years old, HR4796A is believed to be the right age for a system undergoing planetary formation.

Further evidence for proto-planetary disks around young, low-mass stars was provided by the Hubble Space Telescope *images of objects called proplyds, short for protoplanetary disks. This image shows a number of proplyds in the Orion nebula.*

(Image from CSC/STScI and NASA)

Accretion and Fragmentation

The preplanetary clumps grew by accretion from objects that might be imagined to be the size of baseballs and basketballs to *planetesimals*, embryonic *protoplanets* several hundred miles across. The early solar system might have consisted of millions of planetesimals.

Although smaller than mature planets, the planetesimals were large enough to have sufficiently powerful gravitational forces to affect each other. The result must have been a series of near misses and collisions that merged planetesimals into bigger objects, but also fragmentation, as collisions resulted in chunks of some planetesimals breaking off. As you learned in Chapter 6, the formation of the Moon likely happened at this point in the history of the solar system.

The larger planetesimals, with their proportionately stronger gravitational fields, captured the lion's share of the fragments, growing yet larger, while the smaller planetesimals joined other planets or were "tossed out." A small number of fragments escaped capture to become asteroids and comets.

Unlike the planets, whose atmospheres and internal geological activity (volcanism and tectonics) would continue to alter them (Earth, for example, has rocks and minerals that vary greatly in age), asteroids and comets remained geologically static, dead; therefore, their matter, unchanging, marks well the date of the solar system birth.

Star Words

Planetesimals are embryonic planets in an early formative stage. Planetesimals, which are probably the size of small moons, develop into **protoplanets,** immature but full-scale planets. It is the protoplanets that go on to develop into mature planets.

Astro Byte

Astronomers estimate that the evolution from a collection of planetesimals to nine protoplanets, many protomoons, and a protosolar mass at the center of it all consumed about 100 million years. After an additional billion years, it is believed that the leftover materials assumed their present orbits in the asteroid belt, the Kuiper Belt, and the Oort Cloud. The high temperatures close to the protosun drove most of the icy material into the outer solar system where (with the exception of periodic cometary shows) it remains to this day.

Whipping Up the Recipe

Although there is substantial variety among the nine planets, they tend to fall into two broad categories: the large gaseous outer planets, known as the jovians, and the smaller rocky inner planets, the terrestrials (these groups were described in Chapters 8 and 9).

Why this particular differentiation?

As with just about any recipe in any kitchen, part of the difference is caused by heat.

Out of the Frying Pan

As the solar *nebula* contracted and flattened into its pancakelike shape, gravitational energy was released in the form of heat, increasing its temperature. Due to the inverse-square law of gravitational attraction, matter piled up mostly at the center of the collapsing cloud. The density of matter and the temperature were highest near the center of the system, closest to the protosun, and gradually dropped farther out into the disk.

At the very center of the nascent solar system, where heat and density were greatest, the solar mass coalesced. In this very hot region, the carefully assembled interstellar dust was pulled apart into its constituent atoms, while the dust in the outer regions of the disk remained intact. When the gravitational collapse from a cloud to a disk was complete, the temperatures began to fall again, and new dust grains condensed out of the vaporized material toward the center of the solar system. This vaporization and recondensation process was an important step in the formation of the solar system, because it chemically differentiated the dust grains that would go on to form the planets.

Star Words

The term **nebula** has several applications in astronomy and is used most generally to describe any fuzzy patch seen in the sky. Nebulae are often (though not always) vast clouds of dust and gas.

These grains originally had a uniform composition. In the regions nearest the protosun—where temperatures were highest—metallic grains formed, because metals survived the early heat. Moving farther out, silicates (rocky material), which could not survive intact close to the protosun, were condensed from the vapor. Farther out still, there were water ice grains, and, even farther, ammonia ice grains. What is fascinating to realize is that the heat depleted the inner solar system (which is home to Earth) of water ice and organic carbon compounds. These molecules, as you will see, survived in the outer solar system and might have later rained onto the surfaces of the inner planets, making one of them habitable.

The composition of the surviving dust grains determined the type of planet that would form. Farthest from the Sun, the most common substances in the preplanetary dust grains were water vapor, ammonia, and methane, in addition to the elements hydrogen, helium, carbon, nitrogen, and oxygen—which were distributed throughout the solar system. The jovian planets, therefore, formed around mostly icy material. And in the cooler temperatures farthest from the protosolar mass, greater amounts of material were able to condense, so the outer planets tended to be very massive. Their mass was such that, by gravitational force, they accreted hydrogen-rich nebular gases in addition to dust grains. Hydrogen and helium piled onto the outer planets, causing them to contract and heat up. Their central temperatures rose, but never high enough to trigger fusion, the process that produces a star's enormous energy. Thus the jovian worlds are huge, but also gaseous.

Into the Fire

Closer to the protosun, in the hottest regions of the forming solar system, it was the heaviest elements, not ices and gases, that survived to form the planets. Thus the terrestrial planets are rich in the elements silicon, iron, magnesium, and aluminum. The dust grains and then planetesimals from which these planets were formed were rocky rather than icy.

Close Encounter

If the central area of the forming solar system was too hot for light elements and gases to hang around, where did Earth's abundant volatile matter—the oxygen, nitrogen, water, and other elements and molecules so essential to life—come from?

One theory is that comets, which are icy fragments that formed in the outer solar system, were deflected out of their orbits by the intense gravitational force of the giant jovian planets. Bombarded by icy meteoroids after the planet had coalesced and cooled, Earth was resupplied with water and other essential elements.

No matter how they got here, it is fortunate that water ice and organic compounds later rained down on the early Earth, or the present-day planet would be as lifeless as the Moon.

Ashes to Ashes, Dust to Dust

Just as the specifics of the formation of the solar system depended on the formation of the Sun, so its death will be intimately related to the future of our parent star. The evolution of the Sun will presumably follow the same path of other stars of its size and mass, which means that the Sun will eventually consume the store of hydrogen fuel at its core. As this core fuel wanes, the Sun will start to burn fuel in its outer layers, grow brighter, and its outer shell will expand. It will become a red giant, with its outer layers extending as far as the orbit of Venus. When the Sun puffs up into a red giant, Mercury will slow in its orbit, and probably fall into the Sun. Venus and Earth will certainly be transformed, their atmospheres (and, in the case of Earth, also water) being driven away by the intense heat of the swelling Sun. Venus and Earth will return to their infant state, dry and lifeless.

These changes will also have a transforming effect on the outer planets and their moons, warming them significantly. Perhaps even cold, dry Mars will become a more hospitable location.

The question remains, can we extrapolate our understanding of solar system formation to other planetary systems? Clearly some of the details must be the same, but most of the systems that have been discovered are different enough from our own to cause theorists some serious headaches.

Other Planets, Other Worlds

Until fairly recently, the existence of other planetary systems was assumed, but had not been observed. Starting in 1995, though, clear evidence for planets around other stars began to accumulate, and by August of 2003, astronomers had discovered over 100 planetary systems.

Since ancient times, humans have wondered about the possibility of other worlds, perhaps like our own. The Greek philosopher Epicurus in the fourth century B.C.E. contemplated a "plurality of worlds" that existed beyond the scope of our own. His ideas were more akin to some of the arguments for multiple universes that cosmologists make today. Aristotle, in the third century B.C.E., declared the existence of other worlds impossible, because all things like Earth would seek the center (where, obviously, we already were). In the first century C.E., the Roman philosopher Lucretius argued that other worlds like Earth must exist in the vastness of space. The prophet Muhammad weighed in on the topic in the seventh century C.E., commenting that there were many worlds with human inhabitants.

Religious debates about the presence of other worlds continued in Western Europe until 1609, in which Galileo's telescope made it clear that other worlds—at least within the solar system—existed without a doubt. And after other worlds were an accepted fact in the solar system, it wasn't a great leap to think that they might exist around a few of the other billions of stars in the sky.

How to Find a Planet

Models of star formation have made it clear that it would be very difficult to form a star without also forming a planet, a task akin to baking a pie without leaving a speck of flour on the counter. But hard evidence for the existence of planets around other stars eluded astronomers until the eve of the twenty-first century. There is any number of ways to hunt for planetary systems, some more obvious than others. Astronomers have tried every method yet proposed, with varying degrees of success.

Take a Picture

The most obvious approach is just to take a picture. When we want a picture of a planet in the solar system, we simply point a telescope in its direction, snap a photo, and we're done. If we want a really sharp picture, we can send a probe to orbit the planet (like the *Mars Orbiter Camera*). If we want to discover planets around other stars, just point our telescopes in the direction of a suspicious-looking star and *voila!* How hard could it be?

Well, there are a few problems with this approach.

First is the relative brightness of stars and planets. Stars shine because of the nuclear fusion occurring in their cores. Planets shine in the reflected light of their mother star. And planets are much smaller than the stars they are orbiting. As a result, even the largest planet in our solar system (Jupiter) is one billion times fainter than the Sun. To put it bluntly, the "simple" imaging approach is difficult because it's very hard to see a firefly (a planet) circling a multi-megawatt searchlight (a star).

0.8" (14 AU)

Not a planet, but getting close. This picture shows a Gemini *image of the Sunlike star 15 Sge and its newly uncovered brown dwarf companion, 15 Sge B, orbiting at a distance of 14 AU. The image has been processed to remove the light from the primary star in the vicinity of the brown dwarf. Brown dwarfs are larger than jovian planets, but too small to be stars.*

(Image courtesy of NOAO, IfA, M. Liu)

The other problem with imaging directly is that planets and stars are very close to one another relative to how far away they are from us. So our task is to look for something (a planet) very close to something that is very far away (a star). Most Earth-bound telescopes don't have enough resolving power to do this. For the time being, therefore, what would seem the simplest solution is beyond our available technology.

Watching for Wobbles

When a large planet orbits a star, the central star doesn't stay exactly fixed. It moves around a little bit, and that movement betrays the presence of an orbiting planet. The bigger the planet, the more the star moves.

There are two ways to detect star motion. One involves taking images that record the exact positions of the stars. The other involves observing the color spectrum of a candidate star and looking for minute changes in its color with time. Both require meticulous observations and lots of telescope time.

The first technique, called *astrometry*, requires many images of a star and its nearby companions. These images are examined for minute changes in the star's position. If a star is orbited by a particularly large planet, it is possible to measure the wobble in the position of the star. For example, if some slimy alien astronomer were looking at the Sun, he, she, or it would see the Sun wobble back and forth every 11 or 12 years. He, she, or it would be watching the Sun's response to the orbit of Jupiter. All of the other planets would cause wobbles as well, but the wobble from Jupiter would be the dominant one. The problem with this method is that the wobbles caused by even the largest planets are very small, at the limit of even the best telescopes now in operation. (Of course, alien technology might be far more advanced than ours.)

The second technique uses the Doppler effect or Doppler shift that you first encountered in Chapter 4. You hear the Doppler effect when a fire truck races past you, the pitch of the siren going from high to low as the truck passes where you are standing. The sound is "blueshifted" (higher frequency) as the truck approaches, and "redshifted" (lower frequency) as it recedes. Light from a star does the same thing. As an orbiting planet pulls a star away from the viewer, its light is redshifted. When the orbiting planet pulls the star toward the viewer, its light is blueshifted. The pattern of blueshifts and redshifts can be used to determine the size and distance of the planet doing the pulling.

Star Words

Astrometry meticulously analyzes many photographic images taken over time to detect changes in the position of stars. Such changes might reveal the gravitational influence of unseen extrasolar planets orbiting the star under observation.

It should be no surprise that the first systems discovered with this technique were the ones with the biggest planets in the smallest orbits (doing the most pulling), but as more and more systems have been discovered, we are finding systems that resemble our own multiple-planet system. In 1999 the first extrasolar multiple-planet system, Upsilon Andromeda, was discovered with this technique.

Mass Transit

Solar eclipses are regular events on Earth. They occur when the Moon passes in front of the Sun, blocking nearly all of its light. Less obvious events occur when planets like Venus and Mercury pass in front of the Sun, blocking a tiny fraction of its light. These events are called planetary transits, and they happen with a fair degree of regularity, because all of the planets orbit pretty much in the same plane.

However, detecting a planetary transit in a distant star takes both luck and skill. The luck is happening to view another planetary system "edge on." The skill comes in making very sensitive measurements of the brightness of a star. When the brightness dips in a tiny, regular, and cyclical fashion, then a transit has been discovered. The first such detection was made for the star HD 209458 in 1999. The great advantage of this sort of measurement is that the details of the "dip" can reveal the radius of the planet that is transiting, and in some cases, even features of the atmosphere of the orbiting planet.

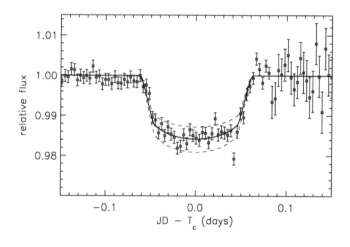

This figure from the discovery paper records the "dip" in the brightness of the star HD 209458 as one of its larger planets passes in front of it (from our perspective). The Kepler mission (discussed later in this chapter) will monitor thousands of nearby stars looking for dips like these.

(Image courtesy of D. Charbonneau)

What Have We Found?

Astronomers were puzzled in the early days of discovering extrasolar planetary systems. Almost none of the systems looked anything like our solar system. The Doppler

Close Encounter

Certainly the most Earthlike system discovered to date is one orbiting the star HD70642 in the constellation Puppis, discovered in 2003 with the Anglo-Australian Telescope (AAT) based in New South Wales, Australia. The new planet, which has a mass of two Jupiters, orbits its star about every six years. The star is only 90 light-years away from Earth.

technique turned up system after system in which a very large (Jupiterlike) planet was in a very small (Mercurylike) orbit. Debates raged, and continue to rage, about how the formation of such systems fits into the model for the formation of our own solar system.

In fact, these discoveries highlight the danger of coming up with a theory of planetary system formation based on one example (our solar system). It is likely that the details of the 100+ planetary and protoplanetary systems that have been discovered will help astronomers improve upon the models that account for the formation of our own system.

Coming Attractions

A number of space missions planned in the coming decade are likely to expand the stable of known planetary systems. Thus far, all the planets discovered have been large and Jupiterlike. Although these discoveries are nothing short of stunning, there is, among some scientists, a burning desire to discover Earthlike planets—terrestrial planets, planets that resemble our own. Are we just searching for ourselves?

Philosophical questions aside, many planned missions will pan for terrestrial gold. Terrestrial Planet Finder (TPF) has as its goal the discovery of terrestrial planets orbiting nearby Sunlike stars. After such planets are pinpointed, the TPF will perform spectroscopy, examining the atmospheres of the planets for evidence of life (the presence of oxygen, for example, or ozone). The launch of this mission is slated for the period 2012–2015. TPF will be preceded by the Space Interferometry Mission (SIM), planned for launch in 2009, to test many of the technologies required for TPF.

And scheduled to launch even earlier, in 2007, the Kepler Mission will scan the skies for planetary transits. It will have instrumentation able to scan thousands of stars at once and to detect very faint transits, the kind that result from planets as small as Earth.

Even now, the Earth-based Keck interferometer (in Hawaii) is taking data. The two giant, 10-m Keck telescopes are being used in conjunction to make images of the dust around nearby stars. Keck observations will also enable astronomers to measure the astrometric wobbles of stars due to Uranus-sized planets—not as small as Earth, but a major step in the right direction.

So stay tuned, we already know that we have a lot of planetary company, and only time will tell what else there is out there.

The Least You Need to Know

◆ Based on studies of meteorites, astronomers believe the solar system is 4.6 billion years old.

◆ The early solar system, like all young planetary systems, was filled with dust grains that were pulled together by gravity to form planetesimals. The composition of the dust grains and thus the planetesimals depended on the distance from the Sun.

◆ A gravitationally contracting nebula in which dust grains condense and collide to form planets can explain most of the patterns observed in our solar system and other known planetary systems.

◆ Stars other than the Sun are host to planetary systems, and some of these systems resemble our own.

◆ There are three main techniques for discovering extrasolar planetary systems; so far, the Doppler effect technique has been the most productive.

Part 3

To the Stars

By far the closest and most familiar star is the Sun, and before we leave the confines of our solar system to explore the nature of stars in general, we take a close look at our own star, at its structure and the mechanism that allows it to churn out, each and every second, the energy equivalent of four trillion trillion 100-watt bulbs.

From the Sun, we move to a chapter that describes how stars are studied and classified. We then explore how stars develop and what happens to stars when they die, including very massive stars that are many times the mass of our Sun. Here, we'll explore the truly strange realm of neutron stars and black holes.

Chapter 12

Our Star

In This Chapter

- ◆ The Sun: an average star
- ◆ Sunspots, prominences, and solar flares
- ◆ The layered structure of the Sun and its atmosphere
- ◆ The Sun as a nuclear fusion reactor
- ◆ Dimensions and energy output of the Sun
- ◆ Something new under the Sun: solar neutrinos

An evening spent looking up under dark skies will convince you that stars can be breathtaking in their loveliness. Yet because of their great distance, theirs is a remote beauty. It's hard to see much similarity between the distant, featureless points of light against a sable sky and the great yellow disk of daytime, whose brilliance overwhelms our vision and warms our world. Yet, of course, our Sun is a star, the very center of our solar system, the parent of the terrestrial and jovian planets and their rings and moons. We are the leftovers of the formation of this burning sphere in the sky.

We spent Part 2 discussing the planets and their moons. But taken together, these objects represent only 0.1 percent of the mass of the solar system. The other 99.9 percent of the mass is found in the Sun. Peoples of many times and cultures have worshipped the Sun as the source of all life. It just makes sense. The Sun is our furnace and our light bulb: the ultimate source of most energy and light here on Earth. And because it contains almost all of the mass, it is the gravitational anchor of the solar system. Indeed, its very matter is ours. The early Sun was the hot center of a swirling disk of gas and dust from which the solar system formed some 4.6 billion years ago. If the Sun were a cake, Earth and the rest of the planets would be nothing more than a bit of flour left on the counter.

Important as it is to us, the Sun is only one star in a galaxy containing hundreds of billions of stars. Actually, astronomers feel fortunate that the Sun is so nondescript a galactic citizen. Its ordinariness lets us generalize about the many stars that lie far beyond our reach. In this chapter, we examine our own star, and begin to explore how the Sun (and stars in general) generate the enormous energies that they do.

The Solar Furnace

The ancient Greek philosopher Anaximenes of Miletus believed the Sun, like other stars, was a great ball of fire. An important insight, but not entirely accurate. The Sun is not quite so simple.

In terms of human experience, the Sun is an unfailing source of energy. Where does all of that energy come from? In the nineteenth century, scientists knew of two possible sources: thermal heat (like a candle burning) and gravitational energy.

The problem with thermal energy is that even the Sun doesn't have enough mass to produce energy the way a candle does—at least, not ongoing for billions of years. Calculations showed that, "burning" chemically, the mass of the Sun would last only a few thousand years. Although a Sun that was a few thousand years old might have pleased some theologians at the time, there was a variety of evidence showing that Earth was much older.

Astro Byte

The description of how mass has an equivalent energy is perhaps the best-known equation of all time: $E=mc^2$. E stands for energy, m for mass (in kg), and c, the speed of light (3×10^8 m/s). Do the math: a tiny bit of mass can produce an enormous

So scientists turned their attention to gravitational energy, that is, the conversion of gravitational energy into heat. The theory went this way: As the Sun condensed out of the solar nebula, its atoms fell inward and collided more frequently as they got more crowded. These higher velocities and collisions converted gravitational energy into heat. Gravitational energy could power the Sun's output at its current rate for about 100 million years.

But when it started to become clear that Earth was much older (geological evidence showed that it was at least 3.5 billion years old), scientists went back to the drawing board, and the nineteenth century ended without an understanding of the source of energy in the Sun.

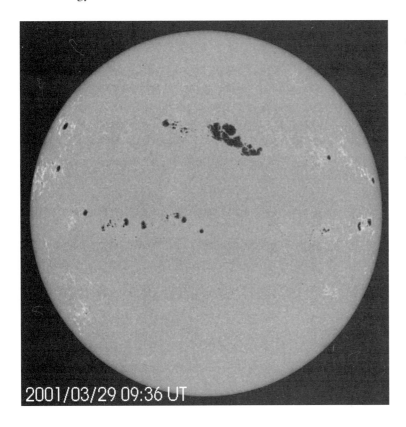

2001/03/29 09:36 UT

The solar surface as seen with the Solar and Heliospheric Observatory (SOHO), *an international solar observatory jointly operated by the European Space Agency and NASA. This image shows an impressive array of sunspots visible on March 29, 2001.*

(Image from NASA/ESA)

A Very Special Theory

At the beginning of the twentieth century, a brilliant physicist named Albert Einstein came up with an answer. It turns out that the source of energy in the Sun is something that had never been considered before—something that would have been called alchemy centuries earlier. It turns out that the Sun converts a tiny bit of its mass into energy through collisions between the cores of atoms, their nuclei.

What's It Made Of?

The Sun is mostly hydrogen (about 73 percent of the total mass) and helium (25 percent). Other elements are found in much smaller amounts, adding up to just under 2 percent of the Sun's mass. These include carbon, nitrogen, oxygen, neon, magnesium, silicon, sulfur, and iron. Over 50 other elements are found in trace amounts. There is nothing unique about the presence of these particular elements; they are the same ones that are distributed throughout the solar system and the universe.

To produce energy, hydrogen atoms in the Sun's core plow into one another and thereby create helium atoms. In the process, a little mass is converted into energy. That little bit of energy for each collision equates to enormous amounts of energy when we count all of the collisions that occur in the core of the Sun. With this energy source, the Sun is expected to last not a thousand years, or even 100 million years, but about 8 to 10 billion years, typical for a star with the Sun's mass.

A Spectacular, Mediocre Star

In terms of its size, mass, and energy released, the Sun is by far the most spectacular body in the solar system. With a radius of 6.96×10^8 m, it is 100 times larger than Earth. Imagine yourself standing in a room with a golf ball. If the golf ball is Earth, the Sun would touch the eight-foot ceiling. With a mass of 1.99×10^{30} kg, the Sun is 300,000 times more massive than Earth. And with a surface temperature of 5,780 K (compared to Earth's average 290 K surface temperature), the Sun's photosphere would melt or vaporize any matter we know.

Four Trillion Trillion Light Bulbs

Next time you are screwing in a light bulb, notice its wattage. A watt is a measure of power, or how much energy is produced or consumed per second. A 100-watt bulb uses 100 joules of energy every second. For comparison, the Sun produces 4×10^{26} watts of power. That's a lot of 100-watt light bulbs—four trillion trillion of them, to be exact. This rate of energy production is called the Sun's *luminosity*. Some stars have luminosities much higher than that of the Sun, and a great many have luminosities that are lower.

The source of the Sun's power—and that of all stars, during most of their life span—is the fusing together of nuclei. Stars first convert hydrogen into helium, and heavier elements come later (we discuss this process in Chapter 15). The only fusion reactions that we have been able to produce on Earth are uncontrolled reactions known as hydrogen bombs. The destructive force of these explosions gives insight into the enormous energies released in the core of the Sun. In the core of the Sun, 600 million metric tons of hydrogen are converted into helium every second. Nuclear fusion could be used as a nearly limitless supply of energy on Earth; however, we are not yet able to create the necessary conditions for controlled fusion reactions.

The Solar Atmosphere

The Sun doesn't have a surface as such. What we call its surface is just the layer that emits the most light. Let's begin our journey at the outer layers of the Sun (the layers that we can actually see), and work our way in. When you look up at the Sun during the day, what you are really looking at is the Sun's photosphere. The layer from which the visible photons that we see arise—the photosphere—has a temperature of about 6,000 K. Lower layers are hidden behind the photosphere, and higher layers are so diffuse and faint (though hot) that we only see them during total solar eclipses or with special satellites. Above the photosphere in the solar atmosphere are the chromosphere, the transition zone, and the corona. As we move higher in the Sun's atmosphere, the temperatures rise dramatically.

Not *That* Kind of Chrome

The part of the Sun's atmosphere called the chromosphere is normally invisible because the photosphere is far brighter. However, during a total solar eclipse, which blots out the photosphere, the chromosphere is visible as a pinkish aura around the solar disk. The strongest emission line in the hydrogen spectrum is red, and the predominance of hydrogen in the chromosphere imparts the pink hue.

The chromosphere is a storm-racked region, into which *spicules*, jets of expelled matter thousands of miles high, intrude.

Star Words

Spicules are jets of matter expelled from the Sun's photosphere region into the chromosphere above it.

Above the chromosphere is the transition zone. As mentioned earlier, the temperature at the surface of the photosphere is 5,780 K, much cooler than the temperatures in the solar interior, which get hotter closer to the core. Yet in the chromosphere, transition zone, and into the corona, the temperature rises sharply the *farther* one goes from the surface of the Sun! At about 6,000 miles (10,000 km) above the photosphere, where the transition zone becomes the corona, temperatures exceed 1,000,000 K. (For detailed real-time views of the solar photosphere, chromosphere, and corona, see http://sohowww.estec.esa.nl.)

How do we explain this apparent paradox? It is believed that the interaction between the Sun's strong magnetic field and the charged particles in the corona heat it to these high temperatures, and its low density makes it more difficult for it to cool.

A Luminous Crown

Corona is Latin for "crown," and it describes the region beyond the transition zone consisting of elements that have been highly ionized (stripped of their electrons) by the tremendous heat in the coronal region. Like the chromosphere, the corona is normally invisible, blotted out by the intense light of the photosphere. It is only during total solar eclipses that the corona becomes visible, at times when the disk of the Moon covers the photosphere and the chromosphere. During such eclipse conditions, the significance of the Latin name becomes readily apparent: The corona appears as a luminous crown surrounding the darkened disk of the Sun. In fact, it is during eclipses, with the corona evident, that the Sun looks most like a child's drawing.

When the Sun is active—a cycle that peaks every 11 years—its surface becomes mottled with sunspots, and great solar flares and prominences send material far above its surface.

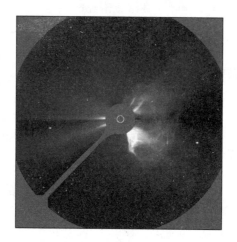

This image was generated by the Large Angle and Spectrometric Coronagraph of SOHO. The solar corona (or crown) can be viewed only if the light from the photosphere is naturally or artificially blocked. The small white circle indicates the size of the solar disk.

(Image from NASA/ESA)

Solar Wind

The Sun doesn't keep its energy to itself. Its energy flows away in the form of electromagnetic radiation and particles. The particles (mostly electrons and protons) don't move nearly as fast as the radiation, which escapes the surface of the Sun at the speed of light, but they move fast nevertheless—at more than 300 miles per second (500 km/s). It is this swiftly moving particle stream that is called the *solar wind*.

The solar wind is driven by the incredible temperatures in the solar corona. As a result, the gases are sufficiently hot to escape the tremendous gravitational pull of the Sun. The surface of Earth is protected from this wind by its magnetosphere, the magnetic "cocoon" generated by the rotation of charged material in Earth's molten core. As with many other planets, the motion of charged molten material in Earth's core generates a magnetic field around the planet.

This magnetic field either deflects or captures charged particles from the solar wind. Some of these particles are trapped in the Van Allen Belts, doughnut-shaped regions around Earth named after their discoverer, American physicist James Alfred Van Allen (1914–). Some of the charged particles rain down on Earth's poles and collide with its atmosphere, giving rise to displays of color and light called *aurora* (in the Northern Hemisphere, the Aurora Borealis, or Northern Lights, and in the Southern Hemisphere, the Aurora Australis, or Southern Lights). The Auroras are especially prominent when the Sun reaches its peak of activity every 11 years. The latest cycle of activity started in 1996, peaked in 2000, and should end in 2006.

> **Astro Byte**
>
> Will the Sun eventually evaporate into solar wind? Every second, a million tons of solar substance is emitted as solar wind. Yet less than a tenth of one percent of the Sun has been thus spent since the beginning of the solar system 4.6 billion years ago.

Fun in the Sun

We have described the layers in the Sun's outer atmosphere, but have ignored some of their more interesting aspects, the storms in the atmosphere. The Sun's atmosphere is regularly disturbed by solar weather in the form of sunspots, prominences, and solar flares. With the proper equipment—or nothing more than an Internet connection (http://sohowww.estec.esa.nl)—you can observe some of the signs of activity on the Sun's surface.

A Granulated Surface

The Sun's surface usually appears featureless, except, perhaps, for sunspots, which we'll discuss in a moment. However, viewed at high-resolution, the surface of the Sun actually appears highly granulated. Now, *granule* is a relative concept when we are talking about a body as big as the Sun. In fact, each granule is about the size of an Earthly continent, appearing and disappearing as a hot bubble of gas rises to the surface of the Sun.

Galileo Sees Spots Before His Eyes

People must have seen sunspots before 1611, when Galileo (and, independently, other astronomers) first reported them. (As recently as March 2001, sunspots easily visible to the unaided eye have appeared.) The largest spots are visible to the naked eye (at

Star Words

Sunspots are irregularly shaped dark areas on the face of the Sun. They appear dark because they are cooler than the surrounding material. They are tied to the presence of magnetic fields at the Sun's surface.

least when the Sun is seen through thin clouds). Yet in Galileo's day, the world was reluctant to accept imperfections on the face of the Sun, and, as far as we know, sunspots were not studied before 1610.

Galileo drew a profound conclusion from the existence and behavior of sunspots. In 1613, he published three letters on sunspots, explaining that their movement across the face of the Sun showed that the Sun rotated.

Sunspots: What They Are

Sunspots are irregularly shaped dark areas on the face of the Sun. They look dark because they are cooler than the surrounding material. The strong local magnetic fields push away some of the hot ionized material rising from lower in the photosphere. A sunspot is not uniformly dark. Its center, called the umbra, is darkest and is surrounded by a lighter penumbra. If you think of them as blemishes on the face of the Sun, just remember that one such blemish might easily be the size of Earth or larger.

Sunspots sometimes persist for months, and they might appear singly, although, usually, they are found in pairs or groups. Such typical groupings are related to the magnetic nature of the sunspots, and the fact that they are correlated with the presence of magnetic "loops." Every pair of spots has a leader and a follower (with respect to the direction of the Sun's rotation), and the leader's magnetic polarity is always the opposite of the follower. That is, if the leader is a north magnetic pole, the follower will be a south magnetic pole.

Sunspot photographed by the National Solar Observatory at Sacramento Peak, Sunspot, New Mexico. Note the small granules and the larger Sunspot umbra, surrounded by the penumbra.

(Image from NSO)

Sunspots are never seen exactly at the equator or near the solar poles, and leaders and followers in one hemisphere of the Sun are almost always opposite in polarity from those across the equator. That is, if all the leaders in the northern hemisphere are south magnetic poles, all the leaders in the southern hemisphere will be north magnetic poles.

We have said that sunspots are thought to be associated with strong local magnetic fields. But why are the fields strong in certain regions of the photosphere?

A meteorologist from Norway, Vilhelm Bjerknes (1862–1951), concluded in 1926 that sunspots are the erupting loops of magnetic field lines, which are distorted by the Sun's differential rotation. That is, like the giant gas jovian worlds, the Sun doesn't rotate as a single, solid unit, but differentially, at different speeds for different latitudes. The Sun spins fastest at its equator, resulting in distortion of the solar magnetic field. The field lines are most distorted at the equator, so that the north-south magnetic field is turned to an east-west orientation. In places where the field is sufficiently distorted, twisted like a knot, the field becomes locally very strong, powerful enough to escape the Sun's gravitational pull. Where this happens, field lines "pop" out of the photosphere, looping through the lower solar atmosphere and forming a sunspot pair at the two places where the field lines pass through the surface into the solar interior.

Sunspot Cycles

In 1843, long before the magnetic nature of sunspots was perceived, astronomer Heinrich Schwabe announced his discovery of a solar cycle, in which the number of spots seen on the Sun reaches a maximum about every 11 years (on average). In 1922, the British astronomer Annie Russel Maunder charted the latitude drift of sunspots during each solar cycle. She found that each cycle begins with the appearance of small spots in the middle latitudes of the Sun, followed by spots appearing progressively closer to the solar equator until the cycle reaches its maximum level of activity. After this point, the number of spots begins to decline. The most recent maximum occurred in early 2001.

Actually, the 11-year period is only half of a 22-year cycle that is more fundamental. Recall that the leading spots on one hemisphere exhibit the same polarity; that is, they are all either north magnetic poles or south (and the followers are the opposite of the leaders). At the end of the first 11 years of the cycle, polarities reverse. That is, if the leaders had north poles in the southern hemisphere, they become, as the second half of the cycle begins, south poles.

Coronal Fireworks

Most frequently at the peak of the sunspot cycle, violent eruptions of gas are ejected from the Sun's surface. The prominences and flares can rise some 60,000 miles (100,000 km) above the photosphere and might be visible for weeks.

> **Astro Byte**
>
> To generate the amount of energy that the Sun does, 600 million metric tons of hydrogen must be fused into helium each and every second. Yet the Sun has so much hydrogen available that it will not exhaust its fusion fuel for at least five billion years.

Solar flares are more sudden and violent events than prominences. Although they are thought also to be the result of magnetic kinks, they do not show the arcing or looping pattern characteristic of prominences. Flares are explosions of incredible power, bringing local temperatures to 100,000,000 K. Whereas prominences release their energy over days or weeks, flares explode in a flash of energy release that lasts a matter of minutes or, perhaps, hours.

At the Core

The Sun is a nuclear fusion reactor. Impressive as its periodic outbursts are, its *steady* production of energy should generate even more wonder—and thought. The Sun has been churning out energy every second of every day for the last four to five billion years.

Gone Fission

On December 2, 1942, Enrico Fermi, an Italian physicist who had fled his fascist-oppressed native land for the United States, withdrew a control rod from an "atomic pile" that had been set up in a squash court beneath the stands of the University of Chicago's Stagg Field. This action initiated the world's first self-sustaining atomic chain reaction. Fermi and his team had invented the nuclear reactor, and the world hasn't been the same since.

Nuclear fission is a nuclear reaction in which an atomic nucleus splits into fragments, thereby releasing energy. In a fission reactor, such as the one Fermi was instrumental in creating, the process of fission is controlled and self-sustaining, so that the splitting of one atom leads to the splitting of others, each fission reaction liberating more energy.

Nuclear fission is capable of liberating a great deal of energy, whether in the form of a controlled sustained chain reaction or in a single great explosion: an atomic bomb. Yet even the powerful fission process cannot account for the tremendous amount of energy the Sun generates so consistently. We must look to another process: *nuclear fusion*.

Star Words

Nuclear fission is a nuclear reaction in which an atomic nucleus is *split* into fragments, thereby releasing energy that previously held the nucleus together. In a fission reactor, the split-off fragments collide with other nuclei, causing them to fragment, until a chain reaction is underway.

Nuclear fusion is a nuclear reaction that produces energy by *joining* atomic nuclei. The mass of a nucleus produced by joining two nuclei is slightly less than that of the sum of the original two nuclei. The mass is not lost, but instead converted into energy.

Whereas nuclear fission liberates energy by splitting atomic nuclei, nuclear fusion produces energy by joining them, combining light atomic nuclei into heavier ones. In the process, the combined mass of two nuclei in a third nucleus is *less* than the total mass of the original two nuclei. The mass is not simply lost, but converted into energy. A *lot* of energy.

One of the by-products of nuclear fusion reactions is a tiny neutral particle called the *neutrino*. It was Fermi who dubbed the particle a neutrino, or "little neutral one" in Italian. The fusion reactions themselves produce high-energy gamma-ray radiation, but on the Sun those photons are converted into mostly visible light by the time their energy reaches the surface of the Sun. Neutrinos, with no charge to slow them down, come streaming straight out of the Sun's core.

Chain Reactions

The Sun generates energy by converting the hydrogen in its core to helium. The details are complex, so we present only the briefest overview. When temperatures and pressures are sufficiently high (temperatures of about 10 million K are required), 4 hydrogen nuclei (which are protons, positively charged particles) can combine to create the nucleus of a helium atom (2 protons and 2 neutrons).

The mass of the helium nucleus created is slightly less than that of the four protons that were needed to create it. That small difference in mass is converted into energy in the fusion process. One of the simplest fusion reactions involves the production of deuterium (a hydrogen isotope) from a proton and a neutron. When these two particles collide with enough velocity, they create a deuterium nucleus (consisting of a proton and a neutron) and the excess energy is given off as a gamma-ray photon. In the Sun, this process proceeds on a massive scale, liberating the energy that lights up our daytime skies. That's a 4×10^{26}-watt light bulb up there, remember?

Your Standard Solar Model

By combining theoretical modeling of the Sun's (unobservable) interior with observations of the energy that the Sun produces, astronomers have come to an agreement on what is called a *standard solar model*, a mathematically based picture of the interior structure of the Sun and its energy-generating "machinery." The model seeks to explain the observable properties of the Sun and also describe properties of its unobservable interior.

With the standard solar model, we can begin to describe some of the interior regions—regions hidden beneath the photosphere, from direct observation. Below the photosphere is the convection zone, which is some 124,000 miles (200,000 km) thick. Below this is the radiation zone 186,000 miles (300,000 km) thick that surrounds a core with a radius of 124,000 miles (200,000 km).

The Sun's core is tremendously dense ($150,000$ kg/m^3) and tremendously hot: some $15,000,000$ K. We can't stick a thermometer in the Sun's core, so how do we know it's that hot? If we look at the energy emerging from the Sun's surface, we can work backward to the conditions that must prevail at the Sun's core. At this density and temperature, nuclear fusion is continuous, with particles always in violent motion. The Sun's core is a giant nuclear fusion reactor.

At the very high temperatures of the core, all matter is completely ionized—stripped of its negatively charged electrons. As a result, photons (packets of electromagnetic energy) move slowly out of the core into the next layer of the Sun's interior, the radiation zone.

Here the temperature is lower, and photons emitted from the core of the Sun interact continuously with the charged particles located there, being absorbed and re-emitted. While the photons remain in the radiation zone, heating it and losing energy, some of their energy escapes into the convection zone, which boils like water on the stovetop so that hot gases rise to the photosphere and cool gases sink back into the convection zone. Convective cells become smaller and smaller, eventually becoming visible as granules at the solar surface. Thus, by convection, huge amounts of energy reach the surface of the Sun. Atoms and molecules in the Sun's photosphere absorb some of the Sun's emerging photons at particular wavelengths, giving rise to the Sun's absorption-line spectrum. Most of the radiation from a star that has the surface temperature of the Sun is emitted in the visible part of the spectrum.

> **Star Words**
>
> Astronomers also use a technique called *helioseismology* to study the Sun's interior. By measuring the vibrational frequencies of various regions on the surface of the Sun, conclusions can be drawn about the Sun's internal structure. This process is analogous to the way in which seismologists monitor and measure the waves generated by earthquakes in order to draw conclusions about the unseen internal structures of our own planet.

The Latest Kink: The Solar Neutrino Problem Solved?

As we suggested previously, neutrinos are produced in the Sun's core as a product of the nuclear fusion reactions occurring there, sometimes called the proton-proton chain.

Unlike photons—which scatter in the solar interior for millions of years before getting to the solar surface—neutrinos interact only weakly with other matter, and therefore stream directly out of the solar interior. Neutrinos, then, give us our only current view of events going on in the core of the Sun.

Theoretical models of the fusion reactions in the solar core predict how many neutrinos we should expect to detect, given the Sun's energy production rate. However, since the very first experiments were run to detect neutrinos from the Sun, scientists have been aware of a problem. None of the experiments detected the number of neutrinos expected based on the standard solar model. There were always too few. The discrepancy varied by experiment, but all came up short by a factor of up to $\frac{1}{3}$. This shortfall was called the Solar Neutrino Problem (SNP), which persisted for 30 years.

In the early 1990s, the GALLEX (GALLium EXperiment) clarified the situation somewhat by detecting lower energy neutrinos than any other detector. These neutrinos resulted from the most common proton fusion reaction, and the results of the experiment made it clear that the shortfall in neutrinos was real, not a problem with the detectors.

So how, then, to explain the Solar Neutrino Problem? There were two main ideas.

First, the core of the Sun might not be as hot as we thought. This would reduce the number of neutrinos that are produced in the core and, therefore, the number detected there.

Second, it was proposed that neutrinos might change ("oscillate") on their journey from the Sun to Earth. There are, in fact, three different types of neutrinos (electron, muon, and tau neutrinos).

The problem seems to have been solved by some recent observations with the Sudbury Neutrino Observatory (SNO), located about 1 mile (2 km) below the surface of Earth in Canada. A joint project of Canada, the United States, and Britain, the SNO in 2001 directly detected some of those changed neutrinos, indicating that the "oscillations" are real. A 30-year-old mystery seems to have been solved.

The Least You Need to Know

- Staggering as the Sun's dimensions and energy output are, the Sun is no more nor less than a very average star.

- The Sun is a complex, layered object with a natural nuclear fusion reactor at its core.

- The gamma rays generated by the fusion reactions in the core of the Sun are converted to optical and infrared radiation by the time the energy emerges from the Sun's photosphere.

- Never look directly at the Sun, especially not through binoculars or a telescope. The safest way to observe the Sun is by projecting its image, either through binoculars or a telescope or simply through a pinhole punched in cardboard.

- Although the Sun has been a dependable source of energy for the last four billion years, its atmosphere is frequently rocked by such disturbances as sunspots, prominences, and solar flares, peaking every 11 years.

- The long-standing "Solar Neutrino Problem" seems to have been resolved. Neutrinos have a tiny amount of mass, and change, or "oscillate," during their journey from the core of the Sun to Earth.

Chapter 3

Of Giants and Dwarfs: Stepping Out into the Stars

In This Chapter

◆ Which stars are nearest and farthest?

◆ Observing and calculating stellar movement

◆ Measuring the size of a star

◆ Classifying stars according to stellar temperatures and chemical composition

◆ Determining the mass of a star

◆ Stellar biographies

In the scheme of things, our own star, the Sun, is relatively accessible. We can make out detailed features on its surface and track the periodic flares and prominences that it emits. But our Sun is only one star of a few 100 billion in our galaxy alone, and all the other stars that exist are much farther away, parts of other distant galaxies.

Earlier, we said that if Earth were a golf ball, the planet Pluto would be a chickpea eight miles away. But even distant Pluto is very close in comparison to our stellar neighbors. On the same scale, the nearest star (in the Alpha Centauri system) would be 50,000 miles away. The other stars are even farther. Much farther.

In this chapter, we begin to reach out to myriad cousins of our Sun.

Sizing Them Up

The planets of the solar system appear to us as disks. When we know a planet's distance from us, it is quite simple to measure the disk and translate that figure into a real measurement of the planet's size. But train your telescope on any star, and all you will see is a point of light, with no disk to measure. Pop in a higher-power eyepiece and guess what? It's still a point of light!

Only very recently (with the advent of the *Hubble Space Telescope*) have we been able to resolve the disks of *any* stars. In 1996, the *HST* took a picture of the star Betelgeuse. This image was the first resolved image of a star other than the Sun. The Very Large Array has also been used to image Betelgeuse. Although far away (500 light-years), Betelgeuse is a red giant, and as you'll soon see, that means it has a very large radius.

The Very Large Array in Socorro, New Mexico, recently made an image of the nearby supergiant star, Betelgeuse, the bright red star in the constellation Orion. Like the Sun, Betelgeuse appears to send plumes of gas far above its optically visible atmosphere, as seen in this radio-frequency image.

(Image from NRAO)

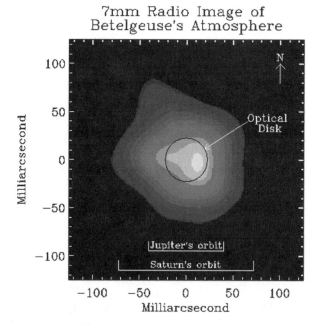

Courtesy of J. Lim, C Carilli, S. M. White, A. J. Beasley, & R. G. Marson

Radius, Luminosity, Temperature: A Key Relationship

We don't have to give up on measuring the sizes of stars, however. We just have to be more clever going about it. What astronomers do is first determine the star's temperature and mass, which can be done by studying its color and spectrum. Then, employing numerical models of how stars hold together, they derive the quantity that they are interested in (radius, for example). It is akin to looking out over a parking lot and seeing a Lexus. Now, you might not know its size, but you know (consulting a chart) that this model of Lexus is 18.5 feet long. You can see clearly that it is indeed this particular model of Lexus, so you know its length, even though you didn't actually measure it with a ruler.

Recall from Chapter 4 that the Stefan-Boltzmann law states that a star's luminosity (its wattage, or the rate at which it emits energy into space) is proportional to the fourth power to the star's surface temperature. This relationship can be extended further. A star's luminosity is not only related to its temperature, but to its surface area. Heat the head of a pin to 400 degrees F and a large metal plate to the same temperature. Which will radiate more energy? Obviously, the object with the larger surface area. Given the same surface temperature, a larger body will always radiate more energy than a smaller one.

This relationship can be expressed in this way: A star's luminosity is proportional to the square of its radius (that's the surface area term) times its surface temperature to the fourth power (Luminosity = $r^2 \times T^4$). Thus, if you know a star's luminosity and temperature (which can be measured by available astronomical instruments), you can calculate its radius. How do you measure a star's luminosity and temperature? Let's see.

The Parallax Principle

First, how do we know which of all the stars in the night sky are the nearest to us? For that matter, how do we know how far away any stars are? You've come a long way in this book, and, on this journey we have spoken a good deal about distances—by Earthly standards, often extraordinary distances. Indeed, the distances astronomers measure are so vast that they use a set of units unique to astronomy. When measuring distances on Earth, meters and kilometers are convenient units. But in the vast spaces between stars and galaxies, such units become inadequate. As you'll see in this chapter and those that follow, the way astronomers measure distances, and the units they use, depends upon how far away the objects are.

Distances between a given point on Earth and many objects in the solar system can be measured with radar. *Radar*, a technology developed before and during World War II, is now quite familiar to most people. Radar can be used to detect and track distant objects by transmitting radio waves, then receiving the returning waves the object reflects back (*sonar* is a similar technique using sound waves). If you multiply the round-trip travel time of the outgoing signal and its incoming echo by the speed of light (which, you'll recall, is the speed of all electromagnetic radiation, including radio waves), you obtain a figure that is twice the distance to the target object.

Star Words

In astronomy, **radar** (short for "radio detection and ranging") is the use of radio signals to measure the distance of planets and other objects in the solar system.

Radar ranging works well with objects that return (bounce back) radio signals. But stars, including the Sun, tend to absorb rather than return electromagnetic radiation transmitted to them.

Moreover, even if you could bounce a signal off the surface of a star, most stars are so distant that you would have to wait hundreds or thousands of years for the signal to make its round trip—even at the speed of light! Even the nearby (relatively speaking) Alpha Centauri system would take about eight years to detect with radar ranging, were it even possible.

Astronomers use another method to determine the distance of the stars, a method that was available long before World War II. In fact, it is at least as old as the Greek geometer Euclid, who lived in the third century B.C.E.

The technique is called *triangulation*—an indirect method of measuring distance derived by geometry using a "baseline" of known size and two angles from the baseline to the object. Triangulation does not require a right triangle, but the establishment of one 90-degree angle does make the calculation of distance a bit easier.

Astronomer's Notebook

If you have a stopwatch, you could also clap your hands and see how long it takes for the echo to return. Using the technique described for radar ranging, but multiplying the round-trip time by the speed of sound in air—about 340 m/s—you get twice the width of the canyon in meters.

It works like this. Suppose you are on one rim of the Grand Canyon and want to measure the distance from where you are standing to a campsite located on the other rim. You can't throw a tape measure across the yawning chasm, so you must measure the distance indirectly. You position yourself directly across from the campsite, mark your position, then turn 90 degrees from the canyon and carefully pace off another point a certain distance from your original position. The distance between your two observation points is called your "baseline."

From this second position, you sight to the campsite. Whereas the angle formed by the baseline and the line of sight at your original position is 90 degrees (you arranged it to be so), the angle formed by the baseline and the line of sight at the second position will be somewhat less than 90 degrees. If you connect the campsite with Point A (your original viewpoint) and the campsite with Point B (the second viewpoint), both of which are joined by the baseline, you will have a right triangle.

Now, you can take this right triangle and, with a little work, calculate the distance across the canyon. If you simply make a drawing of your setup, making sure to draw the angles and lengths that you know to scale, you can measure the distance across the canyon from your drawing. Or if you are good at trigonometry, you can readily use the difference between the angles at Points A and B and the length of the baseline to arrive at the distance to the remote campsite.

An old print of a surveyor using triangulation. The surveyor knows the distance from his viewpoint a to point b (his baseline) and, sighting on distant c, can use the angles ∠cba and ∠cab to work out the distance from a to c.

(Image from arttoday.com)

How Far Away Are the Stars?

Like the campsite separated from you by the Grand Canyon, the stars are not directly accessible to measurement. However, if you can establish two viewpoints along a baseline, you can use triangulation to measure the distance to a given star.

There is just one problem. Take a piece of paper. Draw a line one inch long. This line is the baseline of your triangle. Measure up from that line, say, one inch, and make a point. Now connect the ends of your baseline to that point. You have a nice, normal-looking triangle. But if you place your point several feet from the baseline, then connect the ends of the baseline to it, you will have an extremely long and skinny triangle, with angles that are very difficult to measure accurately, because they will both be close to 90 degrees. If you move your point several *miles* away, and keep a 1-inch baseline, the difference in the angles at Points A and B of your baseline will be just about impossible to measure. They will both seem like right angles.

For practical purposes, a 1-inch baseline is just not long enough to measure distances of a few miles away. Now recall that if our Earth is a golf ball (about 1 inch in diameter), the nearest star, to scale, would be 50,000 miles away. So the baseline created by, say, the rotation of Earth on its axis—which would give 2 points 1 inch away in our model—is not nearly large enough to use triangulation to measure the distance to the nearest stars.

The diameter of Earth is only so wide. How can we extend the baseline to a useful distance?

The solution is to use the fact that our planet not only rotates on its axis, but also orbits the Sun. Observation of the target star is made, say, on February 1, and then is made again on August 1, when Earth has orbited 180 degrees from its position six months earlier. In effect, this motion creates a baseline that is 2 A.U. long—that is, twice the distance from Earth to the Sun. Observations made at these two times (and these two places) will show the target star apparently shifted relative to the even more-distant stars in the background. This shift is called *stellar parallax*, and by measuring it, we can determine the angle relative to the baseline and thereby use triangulation to calculate the star's distance.

Earth's orbit around the Sun can be used to establish an enormous baseline, and to measure the distances to the nearest stars.

(Image courtesy of NASA)

To get a handle on parallax, hold your index finger in front of you, with your arm extended. With one eye open, line up your finger with some vertical feature, say the edge of a window. Now, keeping your finger where it is, look through the other eye. The change in viewpoint makes your finger appear to move with respect to a background object. In our example of stellar parallax, your eyes are the two positions of Earth separated by six months, your finger is a nearby star, and the window edge is a distant background star. This method works as long as the star (your finger) is relatively close—you can experiment to see that the closer your finger to your face, the larger the apparent motion. If the star is too far away, parallax is no longer effective.

Nearest and Farthest

Other than the Sun, the star closest to us is Alpha Centauri, which has the largest known stellar parallax, 0.76 arcseconds. In general, the distance to a star in *parsecs* (abbreviated pc) is equal to 1 divided by the stellar parallax in arcseconds—or conversely, its parallax (in arcseconds) will be equal to 1 divided by the distance in parsecs. The measured parallax, in any case, will be a very small angle (less than an arcsecond).

> **Star Words**
>
> Stellar distances are so great that miles, kilometers, and even astronomical units (A.U.s) quickly become unwieldy. Astronomers use a unit called the **parsec,** which stands for "parallax in arcseconds." One parsec is the distance a star must be from the Sun in order to show a parallax of one arcsecond (1"). That distance is 206,265 A.U. or 10.7×10^{16} feet (3.1×10^{16} meters) or 3.26 light-years.
>
> Another way to think of it is this: A parsec is the distance you would have to be from the

Recall that the Moon takes up about 1,800" (or half a degree) on the sky when full, so the 0.76" parallax measured for Alpha Centauri is $^{0.76}/_{1,800}$, or less than $^1/_{2,000}$ the diameter of the full Moon! Using the previous rule to convert parallax into distance, we find that Alpha Centauri is 1/0.76 = 1.3 pc or 4.2 light-years away. On average, stars in our galaxy are separated by about 7 light-years. So Alpha Centauri is a bit closer than "normal." If a star were 10 pc away, it would have a parallax of $^1/_{10}$ or 0.1".

The farthest stellar distances that can be measured using parallax are about 100 parsecs (about 330 light-years). Stars at this distance have a parallax of $^1/_{100}$" or 0.01". That apparent motion is the smallest that we can measure with our best telescopes. Within our own galaxy, most stars are even farther away than this. As the resolution of earthbound telescopes improves with the addition of adaptive optics, this distance limit will be pushed farther out.

Do Stars Move?

As you saw in Chapter 1, the ancients thought that the stars were embedded in a distant spherical bowl and moved in unison, never changing their relative positions. We know now, of course, that the daily motion of the stars is due to Earth's rotation. Yet the stars move, too; however, their great distance from us makes that movement difficult to perceive, except over very long periods of time. A jet high in the sky, for example, can appear to be moving rather slowly, yet we know that it has to be moving fast just to stay aloft and its apparent slowness is a result of its distance. A jet as far away as a star would not appear to move at all.

Astronomers think of stellar movement in three dimensions:

Star Words

The **proper motion** of a star is determined by measuring the angular displacement of a target star relative to more distant background stars. Measurements are taken over long periods of time, and the result is an **angular velocity** (measured, for example, in arcseconds/year). If the distance to the star is known, this **angular displacement** can be converted into a **transverse velocity**.

◆ The *transverse component* of motion is perpendicular to our line of sight—that is, movement across the sky. This motion can be measured directly.

◆ The *radial component* is stellar movement toward or away from us. This motion must be measured from the Doppler shift apparent in a star's spectrum.

◆ The actual motion of a star is calculated by combining the transverse and radial components.

The transverse component can be measured by carefully comparing photographs of a given piece of the sky taken at different times and measuring the angle of displacement of one star relative to background stars (in arcseconds). This stellar movement is called *proper motion*. A star's distance can be used to translate the angular proper motion into a *transverse velocity* in km/s. In our analogy: If you knew how far away that airplane in the sky was, you could turn its apparently slow movement into a true velocity.

Determining the radial component of a star's motion involves an entirely different process. By studying the spectrum of the target star (which shows the light emitted and absorbed by a star at particular frequencies), astronomers can calculate the star's approaching or receding velocity relative to Earth. Certain elements and molecules show up in a star's spectrum as absorption lines. The frequencies of particular absorption lines are known if the source is at rest, but if the star is moving toward or away from us, the lines will get shifted. A fast-moving star's lines will be shifted more than a slow-moving one's. This phenomenon, more familiar with sound waves, is known as the Doppler effect.

How fast do stars move? And what is the fixed background against which the movement can be measured? For a car, it's easy enough to say that it's moving at 45 miles per hour relative to the road. But there are no freeways in space. Stellar speeds can be given relative to Earth, relative to the Sun, or relative to the center of the Milky Way. Astronomers always have to specify which "reference frame" they are using when they give a velocity. Stars in the solar neighborhood typically move at tens of kilometers per second relative to the Sun.

Artist's conception of the Atacama Large Millimeter Array (ALMA) telescope, now under construction in Chile. The product of a collaboration among the National Radio Astronomy Observatory (NRAO) and a number of European space agencies, including the European Southern Observatory (ESO), the 64 12-meter telescopes will operate at millimeter wavelengths and be movable and rapidly steerable.

(Image courtesy of NRAO/AUI)

de-angle view of the second of four 8.2-m Very
e Telescope (VLT) telescopes (KUEYEN) located
e European Southern Observatory in Chile.

e courtesy of ESO)

Green Bank Telescope (GBT) is located in Green Bank, West
inia, and is the world's largest steerable radio telescope.

e courtesy of NRAO/AUI)

Images of Jupiter and its four Galilean moons, taken by the Galileo space probe, which recently finished its mission to the largest planet in the solar system. True color images reveal the differences in Jupiter's four largest moons.

(Image courtesy of NASA/JPL)

Earth as seen from the Lunar Command Module *while orbiting the Moon. The angular diameter of Earth as seen from the Moon is about 1.9 degrees. Because the Moon as seen from Earth is about half a degree, this indicates that Earth is three to four times the size of the Moon.*

(Image courtesy of NASA)

The last man on the Moon (so far), Eugene Cernan (Apollo 17) stands near the Lunar Rover Vehicle (LRV).

(Image courtesy of NASA)

This image of the bedrock outcropping in Meridiani Planum was made by the panoramic camera of the Mars Exploration Rover Opportunity *before it began detailed exploration.*

(NASA/JPL/Cornell)

Earth as seen by the Shuttle Radar Topography Mission (SRTM). These three images show Earth as viewed from points in space centered above the Americas, Africa, and the western Pacific.

(Image courtesy of NASA/SRTM)

View of Earth as seen from an Apollo spacecraft. Africa and the Arabian Peninsula are obvious in this image, as are storm systems near the South Pole.

(Image courtesy of NASA)

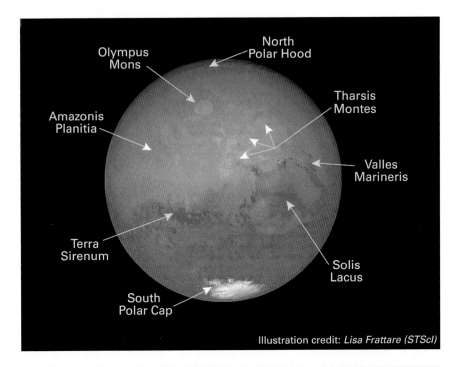

Illustration credit: *Lisa Frattare (STScI)*

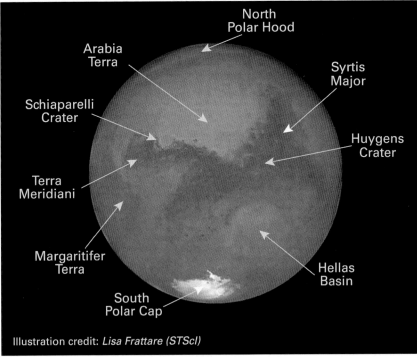

Illustration credit: *Lisa Frattare (STScI)*

These two photos were taken with the Hubble Space Telescope *as Mars made its closest approach to Earth in 60,000 years. The images were taken 11 hours apart and show two opposite hemispheres of Mars. Both hemispheres show most of the southern polar ice cap.*

(Image courtesy of NASA, J. Bell and M. Wolff)

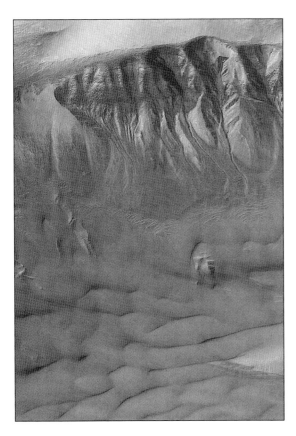

This image of gullies at the rim of a crater (perhaps formed by the release of groundwater) on the surface of Mars were taken by the Mars Orbiter Camera (MOC).

(Image courtesy of NASA/JPL/Malin Space Science Systems)

This microscopic image was taken by the Mars Exploration Rover Opportunity. *The image shows a portion of the bedrock outcropping at Meridiani Planum.*

(NASA/JPL/USGS)

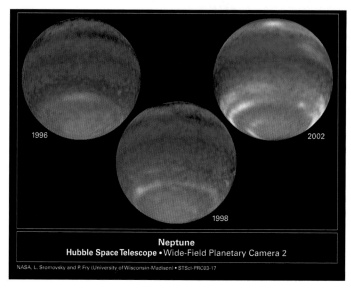

Neptune takes over 164 years to orbit the Sun and experiences seasonal changes due to the tilt of its axis (as does Earth). These images were taken in visible and near-infrared light by Hubble's *Wide Field and Planetary Camera 2.*

(Image courtesy of NASA, L. Sromovsky, and P. Fry)

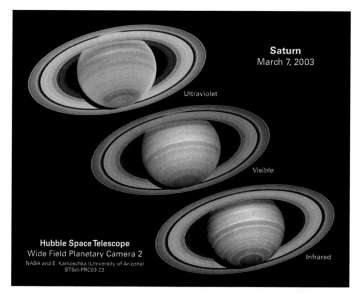

Saturn as seen from its South Pole by the Hubble Space Telescope *in the ultraviolet, optical, and infrared portions of the electromagnetic spectrum. These types of images give insight into the composition and temperature of Saturn's atmosphere.*

(Image courtesy of NASA and E. Karkoschka)

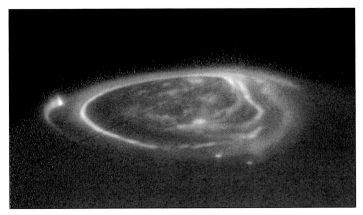

NASA/HST image of the northern aurora of Jupiter. Auroras occur when electrons moving along magnetic field lines crash into the planet's upper atmosphere. On Earth we experience these as the "Northern Lights" and "Southern Lights" of the North and South Poles.

(Image courtesy of NASA/ESA, John Clarke)

Three ultraviolet images indicate the changing level of activity on the Sun's surface. These images show emission from the iron (Fe XII) line at a wavelength of 195 Angstroms. Sunspots, mass ejections, and solar flares all increase at the peak of this 11-year cycle, though in late 2003 an increase in activity was noted outside this normal cycle.

(Image courtesy of EIT/SOHO)

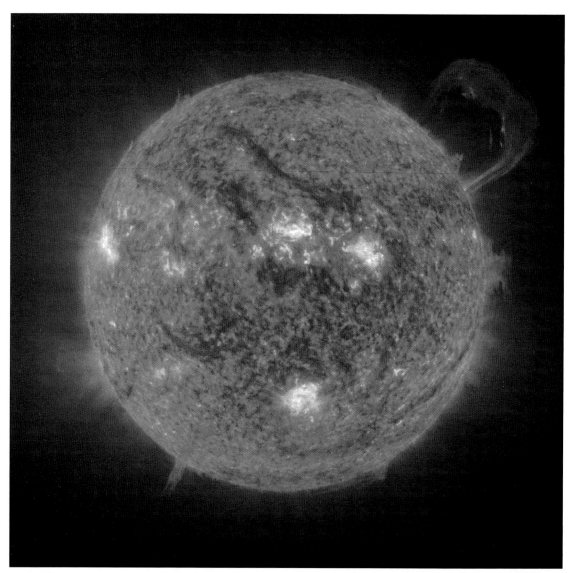

A 304-Angstrom (30.4 nm; ultraviolet) image of a huge solar prominence taken in 1999. Prominences are clouds of cooler plasma suspended in the Sun's corona. All of the details in the image show structure in the magnetic fields. The hottest regions are almost white.

(Image courtesy of EIT/SOHO)

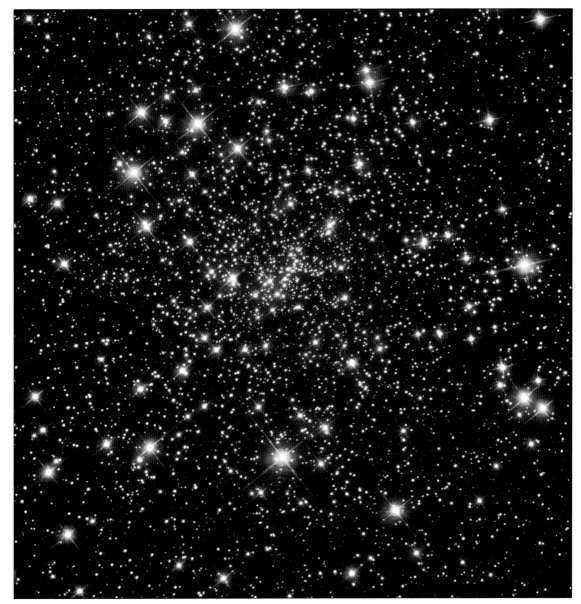

The Hubble Space Telescope *made this image of the core of one of the nearest globular star clusters, NGC 6397. The globular cluster is located 8,200 light-years away from Earth. In globular clusters, the stars are packed in a million times more densely than they are in our Sun's stellar neighborhood.*

(Image courtesy of NASA/AURA/STScI)

The giant galactic nebula NGC 3603 reveals in a single image almost every stage in the birth, life, and death of stars, from the cluster of hot, young stars at the center of the image to the "pillars" of molecular gas at the bottom of the image.

(Image courtesy of Wolfgang Brandner (JPL/IPAC), Eva K. Grebel [Univ. Washington], You-Hua Chu [Univ. Illinois Urbana-Champaign], and NASA)

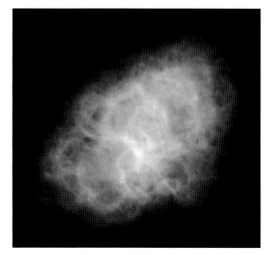

This 6-cm radio image of the Crab Nebula (located in the constellation Taurus) shows emission from the hot material surrounding a pulsar. The remnant consists of material ejected by the star in the supernova explosion and from current pulsar winds.

(Image courtesy of NRAO/AUI and M. Bietenholz)

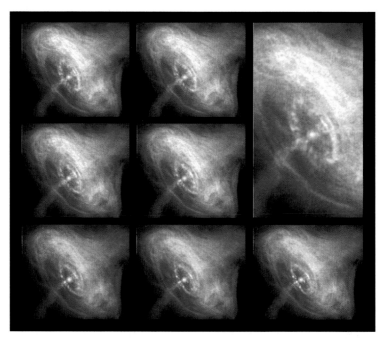

The Crab supernova remnant as imaged with the Chandra Orbiting X-ray Observatory. These images were made over a period of several months. The bright white dot at the center of each image is the pulsar (a rapidly rotating neutron star) that powers the nebula.

(Image courtesy of NASA/CXC/ASU/J.Hester et al.)

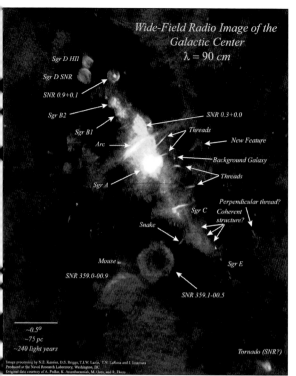

Wide-Field Radio Image of the
Galactic Center
$\lambda = 90$ cm

Sgr D HII
Sgr D SNR
SNR 0.9+0.1
Sgr B2
Sgr B1
Arc
Sgr A
SNR 0.3+0.0
Threads
New Feature
Background Galaxy
Threads
Snake
Sgr C
Perpendicular thread?
Coherent structure?
Mouse
SNR 359.0-00.9
Sgr E
SNR 359.1-00.5
Tornado (SNR?)

~0.5°
~75 pc
~240 light years

Image processing by N.E. Kassim, D.S. Briggs, T.J.W. Lazio, T.N. LaRosa and J. Imamura
Produced at the Naval Research Laboratory, Washington, DC
Original data courtesy of A. Pedlar, K. Anantharamaiah, M. Goss, and R. Ekers

A wide-field image of the center of the Milky Way Galaxy
as seen with the Very Large Array at 90 cm. Notice the
variety of sources, from star-forming regions to supernova
remnants. The bar in the lower left indicates a distance of
240 light-years.

(Image courtesy of NRL/NRAO/AUI and N. Kassim)

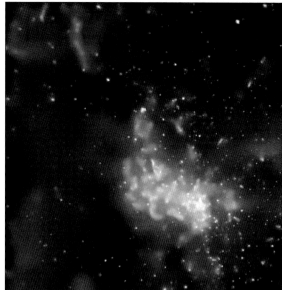

An x-ray image of the supermassive black hole (Sgr A*)
at the center of our own Galaxy, the Milky Way.
Thousands of other background x-ray sources were discovered within this deep image.

(Image courtesy of NASA/CXC/MIT/F.K.Baganoff et al.)

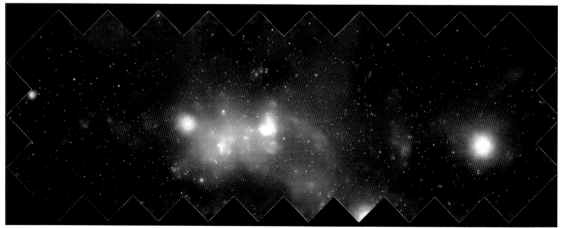

A large (400 by 900 light-year) x-ray mosaic of the central region of our Milky Way Galaxy. Especially apparent in
x-ray emissions are white dwarf stars, neutron stars, and black holes. The colors indicate different x-ray energies: red
(low), green (medium), and blue (high).

(Image courtesy of NASA/UMass/D. Wang et al.)

The distribution of atomic hydrogen at all locations in the sky. These data are from measurements of the 21-cm line of neutral hydrogen from many different radio telescopes, including the NRAO 140-foot and 300-foot telescopes, the 85-foot Hat Creek Telescope, and the 60-foot telescope at the Parkes Radio Observatory in Australia.

(Image courtesy of NRAO/AUI)

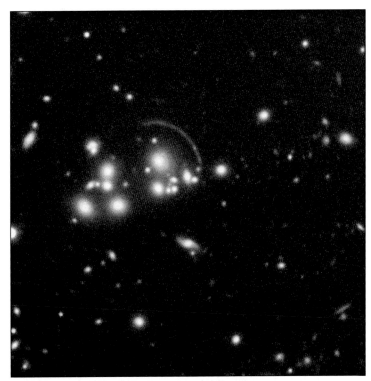

The mass of a foreground cluster of galaxies can distort (or "lens") the light from a distant galaxy, forming gravitational "arcs." This image of gravitational "arcs" in the cluster CL2244-02 was made with one of the VLT 8.2-m telescopes of the European Southern Observatory.

(Image courtesy of ESO)

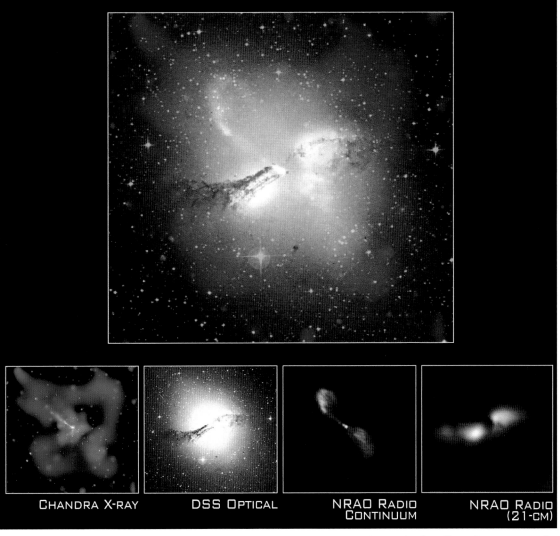

CHANDRA X-RAY   DSS OPTICAL   NRAO RADIO CONTINUUM   NRAO RADIO (21-CM)

Remains of a merger? Astronomers suspect that the current Centaurus A is the result of a collision between a small spiral galaxy and Centaurus A 100 million years ago. A composite x-ray (blue), radio (pink and green), and optical (orange and yellow) image shows the galaxy Centaurus A. Two jets of high-energy material move away from a suspected supermassive black hole in the nucleus.

(Image credits: X-ray image [NASA/CXC/M. Karovska et al.]; radio 21-cm image [NRAO/VLA/J.Van Gorkom/Schminovich et al.]; radio continuum image [NRAO/VLA/J.Condon et al.]; Optical [Digitized Sky Survey U.K. Schmidt Image/STScI])

Messier 100 (one of the largest and brightest spiral galaxies) as imaged with the VLT VIMOS. The VLT Visible Multi-Object Spectrograph (VIMOS)—installed on the 8.2-m PELIPAL telescope—can simultaneously obtain hundreds of spectra of individual galaxies.

(Image courtesy of ESO/VLT)

The spiral galaxy NGC 3370 in its natural environment. Surrounded by a menagerie of galaxies, NGC 3370 was host to a supernova in the early 1990s. The spiral arm structure of the galaxy is peppered with (blue) regions of massive star formation. The galaxy is located over 98 million light-years from Earth.

(Image courtesy of NASA, The Hubble Heritage Team, and A. Riess [STScI])

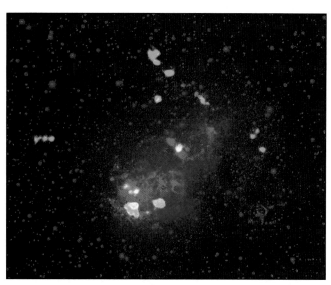

The result of observations with many telescopes, this image shows the dwarf galaxy IC 10 in blue (optical), red (ionized hydrogen), and green (carbon monoxide). Most of its molecular gas (carbon monoxide) is found in the galaxy's outer reaches.

(Image courtesy of OVRO, Caltech, NOAO, KPNO)

When galaxies collide. As two spiral galaxies come together, a wave of star formation is triggered. The large, bright galaxy is NGC 1275 (235 million light years away), and the dusty spiral galaxy in the foreground appears to be sliding through it.

(Image courtesy of NASA and The Hubble Heritage Team [STScI/AURA])

The NASA Hubble Space Telescope *gives us a glimpse of one of the most distant dying stars ever imaged, one that exploded over 10 billion years ago when the universe was young. The results of these types of studies have indicated the presence of "dark energy" in the universe. These images show the supernova's cosmic neighborhood (above), its home galaxy (below left), and the dying star itself (below right).*

(Image courtesy of NASA and A. Riess)

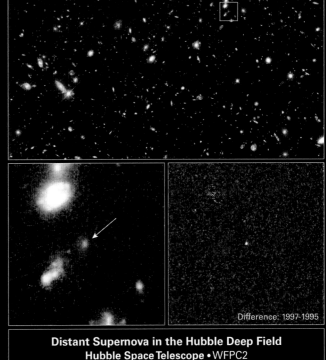

Difference: 1997-1995

Distant Supernova in the Hubble Deep Field
Hubble Space Telescope • WFPC2

NASA and A. Riess (STScI) • STScI-PRC01-09

Hubble Ultra Deep Field
Hubble Space Telescope • Advanced Camera for Surveys

NASA, ESA, S. Beckwith (STScI) and the HUDF Team

STScI-PRC04-07a

This Hubble Space Telescope *"core sample" of the night sky in the Fornax constellation is about* 1/10 *the size of the full moon. Ground-based images of this region show only a few stars in the Milky Way, but this ultra deep image in visible and near-infrared light reveals nearly 10,000 galaxies, the oldest of which formed only 800,000 years after the Big Bang.*

NASA, ESA, S. Beckwith and the HUDF Team

The recently discovered nearby star SO025300.5+165258 has a high transverse angular motion of about 5 arcseconds per year. Bars indicate the changing position of the star with respect to more distant background stars.

(Image courtesy of B. J. Teegarden/NASA/NEAT)

Close Encounter

We have all heard the Doppler effect. It's that change in pitch of a locomotive horn when a fast-moving train passes by. The horn doesn't actually change pitch, but the sound waves of the approaching train are made shorter by the approach of the sound source, whereas the waves of the departing train are made longer by the receding of the sound source.

Electromagnetic radiation behaves in exactly the same way. An approaching source of radiation emits shorter waves relative to the observer than a receding source. Thus the electromagnetic radiation of a source moving toward us will be blueshifted; that is, the wavelength received will be shorter than what is actually emitted. From a source moving away from us, the radiation will be redshifted; we will receive wavelengths longer than those emitted. By measuring the degree of a blueshift or redshift, astronomers can calculate the oncoming or receding velocity (the radial velocity) of a star.

How Bright Is Bright?

In ordinary English, *luminosity* and *brightness* would be nearly synonymous. Not so in astronomy. You are standing beside a quiet road. Your companion, a couple feet away from you, shines a flashlight in your eyes. Just then a car rounds a curve a quarter-mile away. Which is more luminous, the flashlight or the headlights? Which is brighter?

> **Astro Byte**
>
> Does the Sun move? Yes, the Sun and its planets are in orbit around the center of the Milky Way. The solar system orbits the Galaxy about once every 250 million years. Thus, since the Sun formed, the solar system has gone around the merry-go-round of our Galaxy only 15 to 20 times.

Luminosity Versus Apparent Brightness

Ask an astronomer this question, and she will respond that the flashlight, a few feet from your eyes, is apparently brighter than the distant headlights, but that the headlights are more luminous. Luminosity is the total energy radiated by a star each second. Luminosity is a quality intrinsic to the star; brightness might or might not be intrinsic. *Absolute brightness* is another name for luminosity, but *apparent brightness* is the fraction of energy emitted by a star that eventually strikes some surface or detection device (including our eyes). Apparent brightness varies with distance. The farther away an object is, the lower its apparent brightness.

Simply put, a very luminous star that is very far away from Earth can appear much fainter than a less luminous star that is much closer to Earth. Or, a distant, high luminosity star might look brighter in the night sky than myriad closer, but less luminous stars. Thus, although the Sun is the brightest star in the sky, it is by no means the most luminous.

Creating a Scale of Magnitude

So astronomers have learned to be very careful when classifying stars according to *apparent brightness*. Classifying stars according to their magnitude (or brightness) seemed a good idea to Hipparchus (in the second century B.C.E.) when he came up with a 6-degree scale, ranging from 1, the brightest stars, to 6, those just barely visible. Unfortunately, this somewhat cumbersome and awkward system (higher numbers correspond to fainter stars, and the brightest objects—like Venus and the Sun—have negative magnitudes) has persisted to this day.

Hipparchus's scale has been expanded and refined over the years. The intervals between magnitudes have been regularized, so that a difference of 1 in magnitude corresponds to a difference of about 2.5 in brightness. Thus, a magnitude 1 star is $2.5 \times 2.5 \times 2.5 \times 2.5 \times 2.5 = 100$ times brighter than a magnitude 6 star. Because we are no longer limited to viewing the sky with our unaided eyes, and larger apertures collect more light, magnitudes greater than (that is, fainter than) 6 appear on the scale. Objects brighter than the brightest stars may also be included, their magnitudes expressed as negative numbers. Thus the full Moon has a magnitude of –12.5 and the Sun, –26.8.

To make more useful comparisons between stars at varying distances, astronomers differentiate between *apparent magnitude* and *absolute magnitude*, defining the latter, by convention, as an object's apparent magnitude when it is at a distance of 10 parsecs from the observer. This convention cancels out distance as a factor in brightness, and *absolute magnitude* is therefore an intrinsic property of a star.

> **Star Words**
>
> **Magnitude** is a system for classifying stars according to **apparent brightness.** The human eye can detect stars with magnitudes from 1 (the brightest) to 6 (the faintest).

Some key comparative magnitudes are as follows:

◆ Sun: –26.8

◆ Full Moon: –12.5

◆ Venus (at its brightest): –4.4

◆ Vega: 0

◆ Deneb: 1.6

◆ Faintest stars visible to the naked eye: 6

> **Astro Byte**
>
> The *Hubble Space Telescope* is capable of imaging a magnitude 30 star, which has been compared to detecting a firefly at a distance equal to the diameter of Earth.

How Hot Is Hot?

Stars are too distant to stick a thermometer under their tongue. We can't even do that with our own star, the Sun. But you can get a pretty good feel for a star's temperature simply by looking at its color.

As we discussed in Chapter 4, the temperature of a distant object is generally measured by evaluating its apparent brightness at several frequencies in terms of a black-body curve. The wavelength of the peak intensity of the radiation emitted by the object can be used to measure the object's temperature. For example, a hot star (with a surface temperature of about 20,000 K) will peak near the ultraviolet end of the spectrum and will produce a blue visible light. At about 7,000 K, a star will look yellowish-white. A star with a surface temperature of about 6,000 K—such as our Sun—appears yellow. At temperatures as low as 4,000 K, orange predominates, and at 3,000 K, red.

So simply looking at a star's color can tell you a lot about its temperature. A star that looks blue or white has a much higher surface temperature than a star that looks red or yellow.

Stellar Pigeonholes

The color of stars can be used to separate them into rough classes, but the careful classification of stellar types didn't get under way until photographic studies of many spectra were made.

Using the Spectrum

Precise analysis of the absorption lines in a star's spectrum gives us information not only about the star's temperature, but also about its chemical makeup (see Chapter 4).

Star Words

Spectral classification is a system for pigeonholing stars according to surface temperature. The presence or absence of certain spectral lines is used to place stars in a spectral (temperature) class.

Using spectral analysis to gauge surface temperatures with precision, astronomers have developed a system of *spectral classification*, based on the system originally worked out by astronomers at the Harvard College Observatory. The presence or absence of certain spectral lines is tied to the temperatures at which we would expect those lines to exist. The stellar spectral classes and the rough temperature associated with the class are given in the following table.

Spectral Class	Surface Temperature
O (violet)	>28,000 K
B (blue)	10,000–28,000 K
A (blue)	~10,000 K
F (blue/white)	~7,000 K
G (yellow/white)	~6,000 K
K (orange)	~4,000 K
M (red)	<3,500 K

The most massive stars are the hottest, so astronomers refer to the most massive stars they study as "O" and "B" stars. The least massive stars are the coolest. The letter classifications have been further refined by 10 subdivisions, with 0 (zero) the hottest in the range and 9 the coolest. Thus a B5 star is hotter than a B8, and both are hotter than any variety of A star. The Sun is a spectral type G2 star. Our Galaxy and others are chock full of type G2 and lower temperature (less massive) stars.

O6.5	HD 12993
B0	HD 158659
B6	HD 30584
A1	HD 116608
A5	HD 9547
F0	HD 10032
F5	BD 61 0367
G0	HD 28099
G5	HD 70178
K0	HD 23524
K5	SAO 76803
M0	HD 260655
M5	Yale 1755
F4 metal poor	HD 94028
M4.5 emission	SAO 81292
B1 emission	HD 13256

Stars have distinct spectra, which resemble supermarket bar codes in that they encode information about the star from which the spectra came. In this figure, the hottest stars are at the top and the coolest stars at the bottom. Notice the star names often include HD or Henry-Draper catalog designations.

(Image courtesy of KPNO/NOAO/NSF)

From Giants to Dwarfs: Sorting the Stars by Size

The radius of a star can be determined from the luminosity of the star (which can be determined if the distance is known) and its surface temperature (from its spectral type). It turns out that stars fall into several distinct classes. In sorting the stars by size, astronomers use a vocabulary that sounds as if it came from a fairytale:

- ◆ A giant is a star with a radius between 10 and 100 times that of the Sun.

- ◆ A supergiant is a star with a radius more than 100 times that of the Sun. Stars of up to 1,000 solar radii are known.

- ◆ A dwarf star has a radius similar to or smaller than the Sun.

Making the Main Sequence

Working independently, two astronomers, Ejnar Hertsprung (1873–1967) of Denmark and Henry Norris Russell (1877–1957) of the United States, studied the relationship between the luminosity of stars and their surface temperatures. Their work (Hertsprung began about 1911) was built on the classification scheme of a woman from the Harvard College Observatory, Antonia Maury. She first classified stars both by the lines observed and the *width* or *shape* of the lines. Her scheme was an important step toward realizing that stars of the same temperature could have different luminosity. Plotting the relationship between temperature and luminosity graphically (in what is now known as a *Hertzsprung-Russell* diagram or *H-R diagram*), these two men discovered that most stars fall into a well-defined region of the graph. That is, the hotter stars tend to be the most luminous, and the cooler stars are the least luminous.

> **Close Encounter**
>
> Career opportunities for women were limited at the turn of the century. However, women with specialized training in astronomy were able to find employment at the nation's observatories. The Harvard College Observatory first hired women in 1875 to undertake the daunting task of the classification of stellar types. Under the direction of Edward Pickering, the observatory employed 45 women over the next 42 years. Photography and the telescopes of the HCO were beginning to generate vast amounts of astronomical data, in particular, photographic plates filled with individual stellar spectra. It was more economical to pay college-educated women to pore over these data sets than to hire the same number of male astronomers.

The region of the temperature luminosity plot where most stars reside (indicating that they spend the majority of their lifetime there) is called the *main sequence*. Stars that are not on the main sequence are called "giants" or "dwarfs," and you will see in coming chapters how stars leave the main sequence and end up in the far corners of this temperature-luminosity graph.

A Hertzsprung-Russell diagram, or H-R diagram. This plot of a star's temperature, and intrinsic brightness, shows where in its lifespan a particular star is. The solid line (main sequence) is where stars spend the majority of their lifetimes.

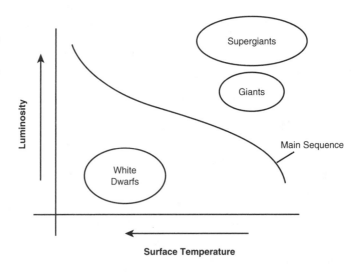

Off the Beaten Track

Although some 90 percent of stars fall into the region plotted as the main sequence, about 10 percent lie outside this range. These include *white dwarfs*, which are far less luminous than we might expect from their high surface temperatures, and *red giants*, which are far more luminous than we would expect from their relatively low surface temperatures. In Chapter 14, we'll describe how stars leave the main sequence and end up as red giants and white dwarfs. Briefly, these are stars that have used up their fuel, and are dying.

Astro Byte

Red dwarfs are the most common stars, probably accounting for about 80 percent of the stellar population in the universe.

Stellar Mass

The overall orderliness of the main sequence suggests that the properties of stars are not random. In fact, a star's exact position on the main sequence and its evolution are functions of only two properties: composition and mass.

Composition can be evaluated if you have a spectrum of the star, its fingerprint. But how can you determine the mass of a star?

Fortunately, most stars don't travel solo, but in pairs known as *binaries*. (Our Sun is an exception to this rule.) *Binary stars* orbit one another.

Some binaries are clearly visible from Earth and are called *visual binaries;* others are so distant that, even with powerful telescopes, they cannot be resolved into two distinct visual objects. Nevertheless, these can be observed by noting the Doppler shifts in their spectral lines as they orbit one another. These binary systems are called *spectroscopic binaries.* Rarely, we are positioned so that the orbit of one star in the binary system periodically brings it in front of its partner. From these *eclipsing binaries* we can monitor the variations of light emitted from the system, thereby gathering information about orbital motion, mass, and even stellar radii.

Star Words

Binaries or **binary stars** are two-star systems, in which the stars orbit one another. The way the companion stars move can tell astronomers much about the individual stars, including their masses.

However we observe the orbital behavior of binaries, the key pieces of information sought are orbital period (how long it takes one star to orbit the other) and the size of the orbit. After these are known, Kepler's Third Law can be used to calculate the combined mass of the binary system.

In 2002, using the *Hubble Space Telescope*, astronomers with the Lowell Observatory discovered the most massive binary star ever detected. The star, contained in the Tarantula nebula in the Large Magellanic Cloud, has a mass almost 60 times larger than the Sun. Why is mass so important?

Mass determines the fate of the star. It sets the star's place along the main sequence, and it also dictates its life span.

The Life Expectancy of a Star

A star dies when it consumes its nuclear fuel, its hydrogen. One might be tempted to conclude that the greater the supply of fuel (the more massive the star), the longer it will live; however, a star's life span is also determined by how rapidly it burns its fuel. The more luminous a star, the more rapid the rate of consumption. Thus, stellar lifetime is directly proportional to stellar mass and inversely proportional to stellar luminosity (how fast it burns). An analogy: A car with a large fuel tank (say a new Hummer H2, which gets 13 mpg) might have a much smaller range than a car with a small fuel tank (a Toyota Celica which gets 29–36 mpg). The key? The Celica gets much better mileage, and thus can go farther with the limited fuel it has.

Thus, while O- and B-type giants are 10 to 20 times more massive than our G-type Sun, their luminosity is *thousands* of times greater. Therefore, these most massive stars live much briefer lives (a few million years at best) than those with less fuel but more modest appetites for it.

A B-type star such as Rigel, 10 times more massive than the Sun and 44,000 times more luminous, will live 20×10^6 years, or 20 million years. For comparison, 65 million years ago, dinosaurs roamed Earth! The Sun (a G2 star) might be expected to burn for $10,000 \times 10^6$ years (10 *billion* years). Our red dwarf neighbor, Proxima Centauri, an M-type star that is $\frac{1}{10}$ the mass of the Sun (and $\frac{1}{100}$ that of Rigel), is only 0.00006 times as luminous as the Sun, and so will consume its modest mass at a much slower rate and might be expected to live longer than the current age of the universe.

The Least You Need to Know

- The distance to nearby stars cannot be measured directly (such as by radar ranging), but it can be determined using stellar parallax. Greater distances can be determined by measuring the period of variable stars.

- Stellar motion, velocity, size, mass, temperature, and luminosity can all be measured. A rough measurement of a star's temperature can be derived from its color. Hotter stars are blue. Cooler stars are red.

- By plotting the relationship between the luminosity and temperature of large numbers of stars, astronomers have noticed that most stars fall along a band in the plot called the main sequence. Stars spend most of their lives on the main sequence.

- The lifetime of a star is determined primarily by its mass. High-mass stars have short lives, low-mass stars have long lives, and all stars eventually die.

Chapter 14

Stellar Careers

In This Chapter

- ◆ Star birth and the interstellar medium
- ◆ Red giants and supergiants
- ◆ From supergiant to planetary nebula
- ◆ The death of a low-mass star
- ◆ The death of a high-mass star
- ◆ Supernovae as creators of elements

The stern Protestant theologian John Calvin would have liked stellar evolution. Each and every star appears "predestined" to follow a certain path in its life, and that path is set only by the mass of the star at its birth. There are a few complications, of course. If stars have nearby companions, they can be "revived" late in life, and astronomers haven't entirely figured out the details of evolution for some types of stars.

But the mass of a star at its birth does unalterably determine where it will reside on what is called the *main sequence*, the period of a star's lifetime devoted to converting hydrogen into helium via nuclear fusion (see Chapter 13 for more on the main sequence). The initial mass determines whether a star will be a relatively cool M-type dwarf star, a hot and massive O or B star, or something in between (like our Sun). When the forces within a star—gravity pulling inward and the force due to the pressure of the heat of fusion pushing outward— reach equilibrium, stars enter the main sequence, their relatively dull middle age. But when the forces get out of balance and stars leave the main sequence to enter their death throes, the fun begins. The final moments of a star can be spectacular, as this chapter will describe.

A Star Is Born

You probably know people with a Type A personality—they are always keyed up, hyper, super-overachievers, and you even call them workaholics. These people often make it big, only to burn out quickly. Then there's the Type B personality: the kind of person who moves through life calmly, doing what's necessary, but no more. These people might not be very spectacular, but as with that tortoise in the fable, slow and steady sometimes wins the race.

Stellar careers show a similar range of "personalities," though the letters of the alphabet used to label them are different. The massive, hot O and B stars are born, mature, and enter the main sequence, only to burn out, their hydrogen fuel exhausted, after a few million years. They are the gas-guzzlers of the galaxy. In contrast, the red dwarfs, stars of type M, low in mass and low in energy output, might not have as much fuel, but they will burn for hundreds of billions of years. Low-mass stars are the economy cars of the stellar world. They are not as flashy as their more massive cousins, but you don't see their steaming hulks by the side of the road, either.

On the Interstellar Median

The space between the stars is filled with two basic components, gas and dust. The vast majority (99 percent) of *interstellar matter* is what we call gas, and the majority of the gas consists of the most abundant element in the universe, hydrogen. The remainder of the material is referred to as "dust," though smoke might be a more accurate description. In fact, you'll see that interstellar dust puts up quite a smoke screen, keeping us from viewing many regions in our own galaxy optically.

The material between the stars is not evenly distributed. Because gas and dust have mass, the material pulls together into clouds and clumps via its own gravity. The patchy distribution of the interstellar medium means that, in some regions of the sky, astronomers can observe objects that are very distant. In other regions, where the interstellar matter is more concentrated, our optical range of vision is more limited. Think of the last time you looked out an airplane window. If the clouds were patchy, there were some directions in which you could see the ground and others in which you could not.

The lowest-density interstellar clouds consist mostly of atomic hydrogen (called HI, an H followed by a roman numeral I and pronounced "H-one"). Until the advent of radio astronomy, this atomic material was impossible to detect. Cooler, higher-density clouds contain molecular hydrogen (H_2).

These regions are the fuel tanks of the universe, ready to produce new stars. The regions closest to young stars are ionized, their electrons stripped away. The 10,000 K gas in these regions (called HII—"H-two"—regions) emits strongly in the optical portion of the spectrum.

Let's look more closely at these different types of matter between the stars.

> **Astronomer's Notebook**
>
> Gas and dust are thinly distributed throughout space and are the matter from which the stars are formed. About 5 percent of our Galaxy's mass is contained in its gas and dust. The remaining 95 percent is in stars.

Blocking Light

How can mere dust block our optical view of the Milky Way? The answer has to do with the size of the dust grains. Think about this for a moment. A satellite dish need not be solid, but can be made out of a wire mesh, perforated by many small holes. This structure does not let the radio waves slip through, like water through a sieve, because radio waves are big and can't slip through. They are so big, in fact, that as long as the holes are small enough, the radio waves don't even know the holes are there. They reflect off the surface of the dish as if it were solid. All electromagnetic radiation (light included) works this way. The waves interact only with things that are about the same size as their wavelength. As luck would have it, optical wavelengths are about the same size as the diameter of a typical dust grain and are therefore absorbed or scattered by dust, even though long-wavelength radio waves pass right through.

The combined effect of the scattering and absorption caused by the dust produces *extinction*. That is, the dust *absorbs* blue light more than red and also *scatters* blue light more than red. As a result, the visible light that makes it through is reddened. _Interstellar reddening_ is increased when objects are farther away, and astronomers have to take this into account when determining the true color of a star.

You might picture interstellar dust as a kind of fog. Certainly, fog, which consists of water molecules and often particulate matter, interferes with the transmission of light, as anyone who has driven in terror along a foggy mountain road can attest.

Close Encounter

If a cloud of dust is between us and a distant star, the light from the star must pass through the dust before it can get to us. Dust allows the longer (redder) wavelengths to pass, but the shorter (bluer) wavelengths are scattered and absorbed. As a result, the light that makes it through is reddened. It is for this reason that the setting Sun looks redder than the Sun at noon. As the Sunlight passes through a thicker slab of the atmosphere near the horizon, its blue frequencies are absorbed and scattered by the atmosphere while its red frequencies pass right through.

Now, interstellar dust is so diffuse that many thousands of miles of it might not obscure anything, but stack billions upon billions of miles of material between you and a star, and it will certainly affect what you see.

Dusty Ingredients

Astronomers have been able to analyze the content of interstellar gas quite accurately by studying spectral absorption lines, the fingerprint elements create by allowing some wavelengths to pass while absorbing others (see Chapter 4). The precise composition of the interstellar dust is less well understood; however, astronomers have some clues.

The 1 percent of interstellar gas that isn't hydrogen or helium contains far less carbon, oxygen, silicon, magnesium, and iron than would be expected based on the amounts of these elements found in our solar system or in the stars themselves. It is believed that the interstellar dust forms out of the interstellar gas, in the process drawing off some of the heavier elements from the gas. So the dust probably contains silicon, carbon, and iron, as well as ice consisting mainly of water with traces of ammonia and methane as well as other compounds. The dust is rich in these substances, while the gas is poor in them.

Flipping Out

Most of the gas in the interstellar medium, some 90 percent, consists of the simplest element, hydrogen. Helium, the second-simplest element, accounts for another 9 percent. The remaining 1 percent consists of other elements. And the gas is mostly cold. Recall that hydrogen consists of a proton and an electron. If there is enough ambient energy, the electron gets bumped up the energy ladder into an *excited state*. But most of these clouds of hydrogen are far from energy sources—stars—and emit no detectable visible light.

But they do emit radio wavelengths.

This image shows an expanding shell of neutral hydrogen in the outer Galaxy. This large shell has a diameter of 360 parsec (more than 1,000 light-years).

(Image courtesy of Canadian Galactic Plane Survey)

In the 1940s, Dutch astronomers were the first to appreciate that radio waves travel unimpeded through clouds of dust, and that hydrogen produces a radio-frequency spectral line. Here's how it happens: Picture both the proton and the electron as spinning like tops. They are either spinning in the same direction or in opposite directions. If they are spinning in the same direction, the hydrogen atom has a little more energy than if they were spinning in opposite directions. Every so often (it takes a few million years), an electron will spontaneously go from the high-energy state to the low-energy state (flip its spin), and the atom will give off a photon with a wavelength of 21 centimeters. This photon travels unimpeded through the Galaxy to our radio telescopes.

Because the 21-cm line (HI line) has a particular frequency, it tells us not only *where* the gas is, but *how* it is moving. As you'll see in the next chapter, the HI line has been invaluable in mapping our own Milky Way galaxy.

Star Light, Star Bright

By definition, there is *nothing* for the amateur astronomer to see with an optical telescope in the regions of the interstellar medium that are cold. The best she can hope for is to see these regions as dark patches in the foreground of bright optical emission. But, closer to stars, where it's hotter, the interstellar medium can be lit up to spectacular effect. Regions of gas that are illuminated by a nearby massive star or stars are called *emission nebulae.*

Star Words

Emission nebulae (singular, **nebula**) are glowing clouds of hot, ionized interstellar gas, located near a young, massive star.

Messier objects M8 and M16 (the Lagoon Nebula and the Eagle Nebula) are two of the more famous hot clouds of interstellar matter, or emission nebulae.

Emission nebulae (also sometimes called HII regions) form around young type O and type B stars. Recall that these stars are tremendously hot, emitting most of their energy in the ultraviolet portion of the spectrum. This radiation ionizes the gas surrounding the star, causing it to glow when the liberated electrons reunite (or recombine) with atomic nuclei. As the electrons cascade down the rungs of the energy ladder, they give off electromagnetic radiation at particular wavelengths. The spectral lines that result can be detected from radio frequencies through the infrared and into the optical and ultraviolet range.

If there are large amounts of gas and dust between you and an emission nebula or HII region, you won't see it optically. The dust grains will absorb or scatter the optical photons from the ionized hydrogen, and you have to resort to high-resolution infrared or radio telescopes. Toward heavily obscured regions of the Milky Way, such as the galactic center, radio and infrared observations have given us almost all the information that we have.

Regions of ionized gas near young stars (HII regions) have a wide variety of sizes. Diameters range from several tens of pc for the so-called giant HII regions, all the way down to tiny HII regions (called compact or ultra-compact HII regions) with diameters as small as $\frac{1}{100}$ of a pc. The smallest HII are deeply embedded in clouds of gas and dust, and thus are only observable at radio and infrared wavelengths.

Radio observations with the Very Large Array have revealed an enormous amount of information about the hot gas located near young, massive stars. Many of these regions are invisible optically but accessible with radio and infrared wavelength observations. This image shows the hot gas in the region that has formed many massive stars called W49A. This region is similar to the Orion Nebula, and is located in our Galaxy.

(Image courtesy of N. Homeier/C. De Pree/NRAO)

A Matter of Perspective

When we *are* able to see part of the interstellar medium visually, it is often because a nearby star has lit it up. But there are also other ways to see. In some directions, a cold dark cloud of gas might fall between Earth and an emission nebula. In such a case, you would see the dark cloud as a black patch against the emission from the ionized gas. In other directions, you would see small patches of the sky where there are few or no stars. It is unlikely that there are truly few stars in these directions. Almost certainly, a dense cloud of gas is in your line of sight.

These gas pillars in the Eagle Nebula, an emission nebula 7,000 light-years from Earth, were imaged by the Hubble Space Telescope. They consist primarily of dense molecular hydrogen gas and dust. This image shows the emission from ionized hydrogen at the surface of the pillars as well as the absorption caused by the gas and dust contained in the pillars.

(Image by Jeff Hester and Paul Scowen of Arizona State University and NASA)

But you don't have to depend on luck. Even if an interstellar cloud of gas doesn't happen to fall between Earth and a background source, there are ways to detect the source. Even at the relatively cold temperatures of these clouds (100 K as compared to the nearly 10,000 K in most emission nebulae), atoms and molecules are in motion. As molecules collide with one another, they are occasionally set spinning about.

It turns out (according to quantum mechanics) that molecules can spin only at very particular rates, like the different speeds of an electric fan, and when a molecule goes from spinning at one rate to spinning at another, it either gives off or absorbs a small amount of energy. We can detect that energy in the form of electromagnetic radiation.

For many molecules, these photons have wavelengths about a millimeter long, which can be detected by special telescopes called "millimeter telescopes" and "millimeter interferometers." With the addition of millimeter telescopes to the astronomic arsenal, there is no portion of the interstellar medium that can escape our notice. With radio telescopes we can image neutral (cold) hydrogen atoms. With optical, infrared, and radio telescopes, we get pictures of the hot gas near young stars. With millimeter telescopes, we can even seek out the cold clouds of gas that contain molecular hydrogen and other molecules.

The Interstellar Medium: One Big Fuel Tank

Why so much fuss over gas and dust? Because, as far as we can tell, this is the raw material from which stars are born. Now, how does a cloud of gas become a star?

Tripping the Switch

A *giant molecular cloud* is subject to a pair of opposing forces. Gravity (as always) tends to pull the matter of the cloud inward, causing it to collapse and coalesce, yet as the constituent atoms of the cloud come together, they heat up, and heat tends to cause expansion, movement outward. Unless some event occurs to upset the balance, the cloud will remain in equilibrium.

We're not sure what causes a molecular cloud to collapse and form stars, but there are several likely possibilities. The expanding shock wave of a nearby supernova explosion might be sufficient to cause a cloud to collapse. Or, as you'll see, a ripple in a galaxy disk called a density wave could also be the trigger. Some astronomers have proposed that a fast-moving star, punching through a molecular cloud, could cause parts of it to collapse to form more stars.

Star Words

Giant molecular clouds are huge collections of cold (10 K to 100 K) gas that contain mostly molecular hydrogen. These clouds also contain other molecules that can be imaged with radio telescopes. The cores of these clouds are often the sites of the most recent star formation.

Whatever the cause, the fact that stars exist makes it clear that some molecular clouds do become gravitationally unstable and begin to collapse. As a cloud collapses, its density and temperature increase, allowing smaller pieces of the cloud (with less mass) to collapse. When a cloud begins to collapse, it breaks into many fragments, which form scores, even hundreds (depending on the original cloud mass) of stars of various masses. The size of each fragment determines the mass of the star that forms.

Letting It All Out

We now join a single fragment in the collapsing cloud that will become a 1-solar-mass star. Over a period of perhaps a million years, the cloud fragment contracts. In this process, most of the gravitational energy released by the contraction escapes into space, because the contracting cloud is insufficiently dense to reabsorb the radiation. At the *center* of the coalescing cloud (where densities are the highest), more of the radiated energy is trapped and the temperature increases. As the cloud fragment continues to contract, photons have a harder and harder time getting out of the increasingly dense material, thereby causing the temperature at the core to rise even higher.

If the original fragment had any slight rotation (as undoubtedly it did), it will be spinning faster now—the spinning cloud is contracting, like a spinning skater pulling in her arms. If the original giant molecular cloud was 10 to 100 parsecs across, then the first cloud fragments will still have been much larger than our solar system. Depending on the eventual stellar type, a cloud fragment on the verge of becoming a protostar might be somewhat smaller than our solar system. This stage of star formation is typically accompanied by dramatic jets of out-flowing material, called Herbig-Haro objects.

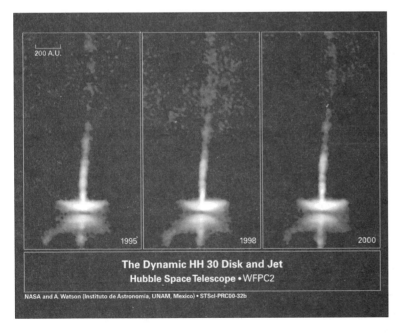

200 A.U.

1995   1998   2000

The Dynamic HH 30 Disk and Jet
Hubble Space Telescope • WFPC2

NASA and A. Watson (Instituto de Astronomía, UNAM, Mexico) • STScI-PRC00-32b

Herbig-Haro objects are jet-like structures that appear to arise during the process of low-mass star formation. High-mass stars might experience very brief, scaled-up versions of these impressive outflows.

(Image from NASA, Alan Watson, Universidad Nacional Autonoma de Mexico, Mexico et al.)

Not Quite a Star

The evolution of a protostar is characterized by a dramatic increase in temperature, especially at the core of the protostar. Still too cool to trigger nuclear fusion reactions, its core reaches a temperature of 1 million K. The protostar is also still very large, about 100 times larger than the Sun. Although its surface temperature at this stage is only half that of the Sun, its area is so much larger that it is about 1,000 times more luminous. At this stage, it has the luminosity and radius of a red giant. A solar mass star will not look like this again until it is on its deathbed, about 10 billion years in the future.

Despite the tremendous heat produced by its continuing collapse, which exerts an outward-directed force on the protostar, the inward-directed gravitational force is still greater.

Through the course of some 10 million years, the protostar's core temperature increases five fold, from 1 million K to 5 million K, while its density greatly increases and its diameter shrinks, from 100 times to 10 times that of the Sun.

Despite the increase in temperature, the protostar becomes less luminous, because its surface area is reduced. Contraction continues but slows as the protostar approaches equilibrium between the inward-directed force of gravity and the outward-directed force of heat energy. It now nears the part of its life called the main sequence.

The "On" Switch

When the star's core temperature reaches about 10 million K, the star begins to fuse hydrogen into helium. The force of gravity is balanced by the pressure of the fusion-produced heat.

Star Words

A **brown dwarf** is a failed star; that is, a star in which the forces of heat and gravity reached equilibrium before the core temperature rose sufficiently to trigger nuclear fusion in the core.

A Collapsed Soufflé

Not all cloud fragments become stars. If a fragment lacks sufficient mass, it will still contract, but its core temperature will never rise sufficiently to ignite nuclear fusion. Failed stars are known as *brown dwarfs*.

In the Delivery Room

Analyzing the process of human gestation and birth is relatively easy. Little theorizing is required because all one needs is about nine months of free time to make some direct observations.

The birth of a low-mass star, however, could consume 40 to 50 million years—obviously more time than any observer can spare. High-mass stars, as you have seen, have shorter lives and more spectacular deaths. They also appear to collapse and fall onto the main sequence more rapidly than low-mass stars. In fact, they live their entire lives in the time it takes a single low mass star to form.

See You On the Main Sequence

After they have matured and assumed their rightful places on the main sequence, stars (regardless of their mass) enjoy a relatively long, stable period in their lives. Indeed, stars in this phase do only one thing in their incredibly hot cores: fuse hydrogen into helium, producing great amounts of energy exactly as we described for the Sun in Chapter 12. This *core hydrogen burning* keeps a star alive or, more precisely, it maintains a star's equilibrium, the balance between the radiation pressure sustained by fusion pushing out and the force of gravity pulling in.

Star Words

Core hydrogen burning is the principal fusion reaction process of a star. The hydrogen at the star's core is fused into helium, producing enormous amounts of energy in the process. This same fusion reaction is what powers a hydrogen bomb.

Running on Empty

From our human perspective, the main sequence life span of an average G-type star such as the Sun seems like an eternity. A good 10 billion years goes by before this kind of star enters the late stages of its life, having fused a substantial amount of its hydrogen into helium.

At this point, the delicate equilibrium between gravity and radiation pressure shifts, and the structure of the star begins to change. At the core of the star is a growing amount of helium "ash." It is referred to as ash (really a misnomer) because it is the end product of the "burning" or fusion of hydrogen. Although hydrogen is fusing in the core, the more massive helium that settles to the star's center cannot reach a sufficient temperature to fuse into a heavier element. As the supply of hydrogen in the core is diluted by the helium ash, the force of gravity starts to win out over the pressure of the slowing fusion reactions. As a result, the core begins to shrink.

Astronomer's Notebook

When a main sequence star becomes a red giant, it swells up to many times its original radius, and becomes far more luminous. When the Sun swells up as a red giant, its luminosity will be about 2,000 times greater than at present, and its radius will be about 150 times greater. That means that the Sun's outer layers will reach out to about 0.7 A.U. (about the orbit of Venus). Earth is at 1 A.U. When this phase happens, Earth's oceans and atmosphere will be evaporated away, leaving only the planet's original rocky surface. But don't lose any sleep—those days are two to five billion years away.

The More Things Change ...

These changes in the core beget other changes farther out. The process of shrinking releases gravitational energy throughout the star's interior. This energy release increases the core temperature, allowing hydrogen that was located outside the star's core to begin fusing while the helium core continues to contract and heat up. At this point, the star's situation is rather paradoxical. Its outer layers are swelling up and shining brightly (called *shell hydrogen burning*), while its core, filling with helium, continues to collapse. It is a short-lived state.

The transition from a G-type star (like the Sun) on the main sequence to the next stage of its career (a red giant) consumes only 100 million years. That's a lot of time if you are human, but it's a blink of the eye relative to the 10-billion-year lifetime of a G-type star—only 1 percent of its total main sequence lifetime.

A Giant Is Born

The helium-enriched core of the aging star shrinks. As it does, the gravitational energy released raises the temperature in the hydrogen-burning shell, increasing the pace of fusion. With this increase, more and more energy is dumped into the outer layers of the star, increasing its temperature (and pressure). So while the core shrinks, the outer layers of the star expand dramatically, cooling as they do. As this happens, the star becomes more luminous (because it's larger) and cooler, moving to the upper right-hand side of the H-R diagram (see Chapter 13). The resulting star is a called a red giant. For a solar mass main sequence star, a *red giant* will be about 100 times larger than the Sun, with a core that is less than 1 percent of the size of the star.

Star Words

A red giant is a late stage in the career of stars about as massive as the Sun. More massive stars in their giant phase are referred to as **red supergiants**. Their relatively low surface temperature produces their red color.

A Flash in the Pan

A red giant continues on its unstable career for a few hundred million years, outwardly expanding and inwardly shrinking. At some point, however, the shrinking of the core raises its temperature high enough to ignite the helium that has been accumulating there patiently. It's like the piston in your car engine on the compression stroke. At a high enough temperature, the combustion reaction is triggered. Only, in a star, it's a nuclear fusion reaction, not chemical combustion, and it requires temperatures in the millions of degrees. Now, along a timeline that has been measured in tens of millions of years, something very sudden occurs. In a process that consumes only a few hours, not millions or billions of years, helium starts to burn in an explosion of activity called the *helium flash*—the explosive onset of helium burning in the core of a red giant star. The helium is now being fused into carbon, and the star settles into another (much shorter-lived) equilibrium.

After the helium flash, the helium in the core is rapidly fused into carbon and oxygen. The star's outer layers shrink, but become hotter, so the star becomes bluer and less luminous. Once helium is exhausted in the star's core, it's nearly at the end of the road for a low-mass star. These stars don't have enough mass to raise the core temperature sufficiently to fuse carbon into heavier elements.

> **Close Encounter**
>
> The night sky offers many examples of red giants. Two of the most impressive are Aldebaran, the brightest star in the constellation Taurus, and Arcturus, in Boötes. Look to the constellation Orion for an example of a red supergiant, Betelgeuse (pronounced *Beetlejuice* by some).

Red Giant *Déjà-Vu*

In astronomical terms—granted, not everyday terms—the equilibrium that results from helium core burning doesn't last long, only some tens of millions of years. Fusion proceeds rapidly, as the star goes about converting helium into carbon and some oxygen. But there's much less helium than there was hydrogen in the star. This fuel runs out much sooner.

Now something very much like the process that created the first red giant phase is replayed, only with helium and carbon instead of hydrogen and helium. As helium is used up in the core of the star, the carbon "ash" settles and the core shrinks yet again, releasing heat. This triggers helium-burning in one shell, which, in turn, is enveloped by a hydrogen-burning shell (left over from the last time this happened). The heat generated in these burning shells causes the star's outer layers to swell, and the star becomes a red giant once again. And when stars return to the red giant phase a second time, they are generally bigger and more luminous than the first time.

We're Losing It

A star might last in this second red giant phase for a mere 100,000 years before its carbon core shrinks to an incredibly dense inert mass about $\frac{1}{1,000}$ the current radius of the Sun—or about the size of Earth. The surrounding shells continue to fuse carbon and helium, and the very outer layers of the star continue to expand and cool.

This planetary nebula in the constellation Ophiucus is known as the "little ghost" as well as NGC 6369. The bright, dying star can be seen in the center of the nebula.

(Image courtesy of NASA/STScI)

The outer layers of the star are now so far from the core that the star's gravity can no longer hold them, and they are able to lift off and move out into interstellar space, often in several distinct shells. These outer layers are sloughed off from the star like spherical smoke rings, leaving behind the bare, hot, carbon-rich core of the star. The cast-off outer layers of the star (which can contain 10 to 20 percent of the mass of the star) are misleadingly called planetary nebulae. Planetary nebulae are the ejected gaseous envelopes of red giant stars. These shells of gas are subsequently illuminated by the ultraviolet photons that escape from the hot, white dwarf star that remains. In truth, a planetary nebula has nothing to do with planets.

We Prefer "Little Stars"

The carbon-rich core of the illuminating star, at the center of the planetary nebula, continues to glow white hot, with surface temperatures of about 100,000 K—much hotter, for example, than the Sun. Nuclear fusion in the star has now ceased; so what keeps it from collapsing further? What is balancing gravity in these stars?

It turns out that a white hot core is so dense that the electrons themselves (which usually have plenty of room to move around freely) are close to occupying the same position with the same velocity at the same time. The laws of nature appear to forbid electrons to do so, and the gas, therefore, can contract no further. Such a gas is called a *degenerate electron gas*, and the pressure of degenerate electrons is all that holds the star up against collapse at this point. The great astrophysicist Chandrasekhar (for whom the Chandra X-ray Observatory is named) first showed that stars could support themselves from further collapse in this way. This white-hot core, about the size of Earth but much more massive (about 50 percent as massive as the Sun), is called a *white dwarf*. One teaspoon of the carbon core would weigh a ton on Earth.

Star Words

A <u>white dwarf</u> is the remnant carbon-rich core of a red giant after it has lost its outer layers as a planetary nebula. Because fusion has now largely stopped, the carbon-oxygen core is supported against further collapse only by the pressure supplied by densely packed electrons. The small size of white dwarfs makes them relatively faint objects, despite their high surface temperatures.

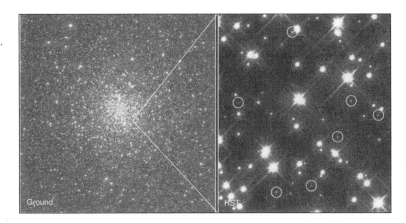

This Hubble Space Telescope *image (right) reveals a population of faint white dwarfs (the circled stars) in globular cluster M4 (Messier object number 4), in the direction of Scorpius. The panel on the left is a view of this region of the sky from a ground-based telescope.*

(Image by Harvey Richer of the University of British Columbia and NASA)

As the white dwarf continues to cool, radiating its stored energy, it changes color from white to yellow to red. Ultimately, when it has no more heat to radiate, it is theorized that it will become a black dwarf, a dead, inert ember, saved from gravitational collapse by the resistance of electrons to being compressed beyond a certain point. Some astronomers have proposed that the carbon atoms will eventually assume a lattice structure. The stellar corpse might become, in effect, an enormous diamond with half the mass of the Sun.

What's New?

Nova is the Latin word for "new," and it seemed to early astronomers an appropriate term for stars that suddenly appeared in the sky. They were obviously *new* stars. Well, they were new in that they hadn't been observed before. But, in fact, a nova is a phenomenon associated with a very *old* star.

When a binary pair of stars (see Chapter 13) forms, it is unlikely that they will have the same mass. And because the mass of a star determines its lifetime (as we have said), the stars in a binary pair will (more often than not) be at different points in their evolution. Sometimes, when a white dwarf is part of a binary system in which the companion star is still youthful (on the main sequence or in a red giant phase), the dwarf's gravitational field will pull hydrogen and helium from the outer layers of its companion. This gas accumulates on the white dwarf, becoming hotter and denser until it reaches a temperature sufficient to ignite the fusion of hydrogen into helium. The flare-up is brief—a matter of days, weeks, or months—but, to Earthly observers, spectacular.

Given the right circumstances, a white dwarf in a binary pair can go nova repeatedly, reigniting each time enough material is stolen from its companion star.

The binary system XZ Tauri consists of two stars separated by approximately the distance between the Sun and Pluto. Images taken over the course of five years reveal that the binary system is associated with an enormous expanding bubble of hot gas. The binary stars are the two bright dots in the lower part of each of the three frames.

(Image courtesy of NASA and J. Krist, STScI)

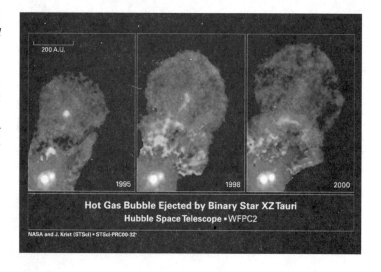

200 A.U.

1995   1998   2000

Hot Gas Bubble Ejected by Binary Star XZ Tauri
Hubble Space Telescope • WFPC2

NASA and J. Krist (STScI) • STScI-PRC00-32

The Life and Death of a High-Mass Star

Up to this point, we've concentrated on the life of a star like the Sun. In truth, most of what we have described (up to the planetary nebula phase) also applies to a high-mass star. It has already been mentioned that stars of significantly lower mass than our Sun enter the main sequence and stay there not for millions or billions of years, but much longer. Stars of significantly greater mass than the Sun (somewhere between 5 and 10 solar masses) have a very different, and more dramatic, destiny.

Fusion Beyond Carbon

Astronomers generally think of 5 to 10 solar masses as the dividing line between low- and high-mass stars. That is, stars about 5 to 10 times more massive than the Sun die in a way very different from stars of lesser mass. The major difference in their evolution is that high-mass stars are able not only to fuse hydrogen, helium, carbon, and oxygen, but also heavier elements in their cores. As core burning of one element ends, it begins burning in a shell outside the core, resulting in a nested series of shells by the time a high-mass star reaches its end.

Last Stop: Iron

Hydrogen fusion produces helium ash that settles to the star's core. Then helium fusion produces carbon ash, which again settles to the star's core. As the fusion of each heavier element proceeds, the core layers progressively contract, producing higher and higher temperatures. Unlike low-mass stars, in a high-mass star the gravitational forces are sufficient to drive core temperatures high enough to fuse carbon into oxygen, oxygen into neon, neon into magnesium, and magnesium into silicon. The end of the road, however, is iron. When a massive star has iron building up in its core, the grand finale is near. The reason is that for every element up until iron, energy was released when nuclei were fused. But to fuse iron into heavier elements absorbs energy. In terms of fusion, iron is a dead end. And to get beyond iron, the universe needs a large input of energy. Because the periodic table does not stop with iron, you can guess that high-mass stars somehow provide the required energy.

The evolution of the high-mass star is rapid. Its hydrogen burns for 1 to 10 million years, its helium for less than 1 million years, its core of carbon for a mere 1,000 years, oxygen for no more than a year, and the fusion of silicon consumes only a week. An iron core grows as a result, but for less than a day. Just before its spectacular death, a massive star consists of nested shells of heavier elements within lighter elements, all the way down to its iron core.

Over the Edge

At this point, the core of a high-mass star is, in effect, an iron-rich white dwarf. It is supported by its degenerate electrons. But there is a problem. The mass of a high-mass stellar remnant is so large that gravity overwhelms even the resistance of electrons to having the same position and velocity. A mass of 1.4 solar masses is sufficient to overwhelm those electrons, which are fused with protons to create neutrons. The temperatures in the core of the star become so high that all of the work of fusion is rapidly undone. The iron nuclei are split into their component protons and neutrons in a process called photodisintegration.

As the core of the star collapses under its own gravity, the electrons combine with protons to become neutrons and neutrinos. The neutrinos (largely unimpeded) escape into space, heralds of disaster. The core of the star only stops its collapse when the entire stellar core has the density of an atomic nucleus. This sudden halt in the collapse causes a shock wave to move through the outer layers of the star and violently blow off its outer layers.

Supernova: So Long, and Thanks for All the Fusion

"Violently" is an understatement. The process of evolution from a hydrogen-burning star to a collapsing core consumed 1 to 10 million years. The final collapse of a high-mass core takes less than a second and ends in a *core-collapse supernova*.

Star Words

Core-collapse supernova is the extraordinarily energetic explosion that results when the core of a high-mass star collapses under its own gravity.

You Know the Type

Two types of supernovae are recognized. Type I supernovae contain little hydrogen, whereas Type II are rich in hydrogen. Only Type II supernovae are associated with the core collapse of high-mass stars. Type I supernovae are associated with our friends the white dwarfs.

Supernovae as Engines of Creation

As you might expect, an explosion as tremendous as that of a supernova creates a great deal of debris. The Crab Nebula, in the constellation Taurus, is the remnant of a supernova that appeared in 1054 C.E. Chinese astronomers left records of that event, reporting a star so brilliant that it was visible for a month in broad daylight. The bright radio source Cassiopeia A is also a supernova remnant.

Close Encounter

No supernova has appeared in our galaxy, the Milky Way, since 1604. Theory predicts a supernova occurrence in our galaxy every 100 years or so. We are, therefore, a bit overdue, and we might be in for a spectacular display any day now. The cosmic rays and electromagnetic radiation that would rain down on Earth if a nearby Type II supernova were to go off (say within 30–50 light-years) could have results that would make a massive asteroid impact look like fun. Fortunately, there aren't any stars that close to us massive enough to generate a Type II supernova.

Amateur astronomers have been credited with many supernova discoveries. For an impressive example of supernova discoveries by Puckett Observatory (65 as of October 2003), see www.astronomyatlanta.com/nova.html.

But supernovae create more than glowing remnants.

Hydrogen and helium, the two most basic elements in the universe, are also the most primitive. They existed before the creation of the stars. A few other elements (carbon, oxygen, neon, silicon, and sulfur) are created by nuclear fusion in low- and high-mass stars. But many of the elements critical to life in the universe are created only in supernova explosions. Only in these explosions is there enough energy to bring nuclei together with sufficient force to create elements heavier than iron.

The only elements that existed at the beginning of the universe were hydrogen and a little bit of helium, beryllium, and lithium. The rest of the periodic table was generated by stars. Each one of us contains the debris of a supernova explosion.

Discovery *image of Type II Supernova 2003J. Horizontal bars indicate position of the supernova.*

(Image from T. Puckett and M. Peoples)

Leftovers

What could possibly survive a supernova? A supernova explosion pushes things out, away from the star's core. But, depending on the original mass of the star, there will be something left behind. Either a bizarre object known as a *neutron star* or, even stranger, a *black hole*.

And You Thought Your Roommate Was Dense

Although the explosive shock wave originates in the core, it doesn't start at the core's very center. If the core's mass was between 1.4 and 3 solar masses, the remnant at the center will be a ball of neutrons known as a neutron star.

In size, this so-called *neutron star* is small by astronomical standards—and tiny, compared to the high-mass star (many times the radius of our Sun) of which it was a part. The neutron star's diameter would be something over 12 miles (20 km), and its density a staggering 10^{17} kg/m^3. All of humanity (if compressed to the density of a neutron star) would be the size of a pea.

The Stars Are Spinning

Neutron stars don't stand still. Just like the stars of which they are a remnant, they rotate. And because they have collapsed from a much larger size, they spin very rapidly. Remember, thanks to conservation of angular momentum, a rotating body (like a whirling skater drawing in her arms) spins faster as it shrinks. A neutron star has shrunk from a body hundreds of times larger than the Sun to one that is smaller than Earth. Earth takes 24 hours to make one revolution. A very massive but very small neutron star rotates in a fraction of a second.

The rate of rotation isn't the only property that intensifies in the neutron star. Its magnetic field is many times stronger than that of the parent star because the lines of the magnetic field are compressed along with the matter of the core itself. The combination of rapid rotation and a powerful magnetic field serves to announce the presence of some neutron stars in the universe.

Star Words

A **neutron star** is the super-dense, compact remnant of a massive star, one possible survivor of a supernova explosion. It is supported by degenerate neutron pressure, not fusion. The nuclei of the atoms in the star cannot be compressed any further. It is an entire star with the density of an atomic nucleus.

In the late 1960s, S. Jocelyn Bell Burnell was a graduate student at Cambridge University working with Anthony Hewish looking for interesting sources of radio emission. One very strange signal was detected: a short burst of radio emission followed by a brief pause, and then another pulse. The pulses and pauses alternated with great precision—as it turns out, with a precision greater than that of the most advanced and accurate timepieces in the world.

In 1974, Hewish alone received the Nobel Prize in Physics for the discovery of the radio signals now called *pulsars*.

A Stellar Lighthouse

Imagine the pulsar as a stellar lighthouse. At the magnetic poles of the neutron star (though not necessarily aligned with the star's rotational axis) are regions in which charged particles are accelerated by the star's magnetic field. These regions, which rotate with the star, radiate intense energy. As the neutron star rotates, a beam of electromagnetic radiation (especially intense in the radio regime) sweeps a path through space. If Earth lies in that path, we see the pulsar.

Thus, all pulsars are neutron stars, but not all neutron stars will necessarily be pulsars (from our vantage point). If the beam of a particular neutron star doesn't sweep past Earth, we will not detect its radio pulsations. Pulsar periods can range anywhere from milliseconds to a few seconds.

Star Words

A **pulsar** is a rapidly rotating neutron star whose magnetic field is oriented such that it sweeps across Earth with a regular period.

Astro Byte

When pulsars were first detected in the late 1960s, their signals were so regular that some astronomers thought they might be a sign of extraterrestrial intelligence. However, the large number of pulsars detected and the neutron star theory of pulsars provided an adequate and simpler explanation.

Morbid Obesity

In the next chapter, we explore the nature of an even more strange supernova leftover: a black hole. If the core of the star is more massive than three solar masses, not even neutron degeneracy pressure—the fact that the neutrons cannot be compressed further—can support it. In this case, the gravitational collapse continues with nothing to halt it, and a black hole is born.

The Least You Need to Know

◆ A star's mass is the primary determinant of the course its life will take. High-mass stars are short-lived (tens of millions of years), while low-mass stars are long-lived (tens of billions of years).

◆ When a star of any mass has fused most of the hydrogen in its core, its days are numbered.

◆ Low-mass stars evolve into red giants and, ultimately, into white dwarfs and planetary nebulae. Although some white dwarfs in binary systems can periodically reignite as novae, white dwarfs gradually cool, finally becoming burned-out embers of carbon and oxygen.

◆ High-mass stars (stars greater than 5–10 solar masses) die spectacularly when their cores collapse, creating a supernova. The remnant is either a neutron star or a black hole.

◆ Many of the elements important to life on our planet in the periodic table are produced only in supernova explosions.

Black Holes: One-Way Tickets to Eternity

In This Chapter

- ◆ What are black holes?
- ◆ Relativity theory
- ◆ Seeing black holes
- ◆ Recent black hole discoveries

Astronomy requires us to contemplate distance scales and time spans far beyond our everyday experience. Can we really fathom our Sun swelling up to the size of the orbit of Venus, or the power of a supernova explosion, briefly bright enough to outshine its host galaxy? Or an entire star ripping itself to shreds just because it ran out of fuel?

If you thought that understanding the lives of stars made you stretch your mind, get ready for even more strenuous mental calisthenics. We have saved some of the strangest objects in the universe for this chapter.

We now explore the end state of a massive star: a black hole.

When the iron core of a massive star is collapsing, it might stop when the entire core of the star has the density of an atomic nucleus, making it a neutron star. If the core is massive enough (more than three solar masses), the collapse becomes unstoppable and overwhelming.

Under Pressure

In the case of ordinary stars, equilibrium is reached when the outward-directed forces of heat (from fusion reactions) are in balance with the inward-directed forces of gravity. Neutron stars, however, produce no new energy, and instead radiate away the heat that is stored in them. They resist the crush of gravity not with the countervailing radiation pressure from fusion, but with neutrons so densely packed that they simply cannot be squeezed any more. Astronomers call these *degenerate neutrons*. They are like passengers in a Tokyo subway car, jammed in by truncheon-toting subway monitors until there is room for no more.

If this is the case, the neutron star isn't so much in equilibrium as it is in stasis. A stalemate exists between the irresistible force of gravity and immovable objects in the form of a supremely dense ball of neutrons.

But if the force of gravity is sufficient, the collapse is apparently unstoppable.

The Livin' End

Incredible though it seems, if a star is massive enough it will continue to collapse on itself. Forever. It's as if those Tokyo subway passengers all suddenly fall into a point in the center of the car. And keep falling. Forever.

Recall from the preceding chapter that a white dwarf evolves from a low-mass parent star (a star less than 5 to 10 solar masses) and that the resulting white dwarf can be no more massive than 1.4 solar masses. If it has a higher mass, gravity will overwhelm the tightly packed degenerate electron pressure, and the core will continue to collapse. When a star's mass is greater than 1.4 solar masses, its core collapse continues, and it will blow off its outer layers as a supernova. If the mass of the remaining core is greater than 1.4 but less than 3 solar masses, the remnant will be a neutron star. However, the "specs" for a neutron star also have an *upper* mass limit. It is believed that a neutron star can be no more massive than about three times the mass of the Sun. Beyond this point, its core of neutrons will yield to gravity's pull.

Astronomer's Notebook

The parent star of a collapsing neutron star core that is more than 3 solar masses would have to be 20 to 30 times more massive than the Sun. In other words, only main sequence stars with masses upwards of 20 solar masses will ever collapse into black holes. Which stars are they? These are a small subset of the massive, hot O and B stars that we have mentioned. We'll study the nurseries of these stars in the next chapter.

X-ray image of the super-nova remnant in Cassiopeia. The exploded remnants of dying massive stars persist for centuries.

(Image courtesy of Chandra X-Ray Observatory)

What results from this apparent stalemate between the force of gravity and the incompressibility of neutrons? Do we just get an even smaller and denser neutron star?

Not at all. We get something completely different, an object from which there is literally no escape.

When an extremely massive star is ripped apart in a supernova explosion, it might produce a supernova remnant so massive that the subsequent core collapse cannot be stopped. When this happens, *nothing* escapes the attractive forces near the core—not even electromagnetic radiation, including visible light. It is called a *black hole*.

Star Words

A stellar mass **black hole** is the end-result of the core collapse of a high mass star. It is an object from which no light can escape. Although space behaves strangely very close to a black hole, at astronomical distances the black hole's only effect is gravitational. If the Sun were replaced with a black hole of the same mass, the planets in the solar system would orbit just as they do now.

No Escape

Although light—and thus information—cannot escape from a black hole, the black hole has certainly not ceased to exist. It is still a physical object with mass. That is the reason it creates a gravitational field, just as Earth or any other object with mass does. But how do we talk about the size of a black hole when we have just described it collapsing without end? Is it infinitely small?

Astro Byte

If the Earth were compressed to the size of a garbanzo bean—yet retained its current mass—a velocity greater than the speed of light would be required to escape its pea-sized surface. Such an object would be, effectively, a black hole. Black holes are "black" because nothing, not even light, can escape their gravitational pull.

Star Words

Escape velocity is the velocity necessary for an object to escape the gravitational pull of another object.

Well, in fact, it does collapse to a point called a "singularity"; however, there is also a dimension associated with black holes. We can talk about the mass of a black hole as well as its radius, but it is a very different kind of radius, as you'll see in a moment. First, let's take a quick detour.

Building a rocket capable of escaping Earth's gravitational pull requires an engine capable of delivering sufficient thrust to achieve a velocity of about 7 miles per second (11 km/s). This *escape velocity* depends upon two factors: the mass of the planet and its radius. For a fixed-mass object, the smaller the radius, the greater the escape velocity required to get free of its gravitational field.

So as the core of a star collapses (with its mass remaining constant), the escape velocity from its surface increases rapidly. You might now wonder: What about when a star has no more "surface" that we can talk about? What is the surface of a black hole? And is there a limit to the increase in this escape velocity?

Nature has one very strict speed limit: the speed of light. Nothing in the universe, not even photons carrying information from distant reaches of the universe, can move faster than the speed of light: 984,000,000 feet per second (300,000,000 m/s). This is the upper limit to escape velocity. When a body of a given mass reaches a certain—very small—size, objects would have to be moving faster than the speed of light to escape. Because nature does not allow this, escape is not an option.

What's That on the Horizon?

But how can we talk about the "surface" of a black hole? A black hole has collapsed to a point of infinite density, and a point has no surface.

The German astronomer Karl Schwarzschild (1873–1916) first calculated what is now called the *Schwarzschild radius* of a black hole. For a star of a given mass, this value is the radius at which escape velocity would equal the speed of light (and, therefore, the radius within which escape is impossible). For Earth, the Schwarzschild radius is the size of a garbanzo bean, about 0.4 inches, or 1 centimeter. For a neutron star at three solar masses, the Schwarzschild radius is 5.58 miles (9 km). So this radius doesn't define a literal "surface" so much as a characteristic property of a black hole.

Remember that the collapse of a black hole is in some sense infinite. Our three-or-more solar-mass stellar core will not stop shrinking just because it has reached the Schwarzschild radius. It keeps collapsing. When it is smaller than the Schwarzschild radius, however, it will effectively disappear. Its electromagnetic radiation (and the information that it carries) is thereafter unable to escape. We spoke earlier about electromagnetic radiation carrying energy and information out into the universe. Because we cannot receive radiation from within the Schwarzschild radius, we cannot get any information from there, either. Events that occur within that radius are hidden from our view. For this reason, the Schwarzschild radius is also called the *event horizon*. We cannot see past it.

Star Words

The **Schwarzschild radius** of a black hole is the radius of an object with a given mass at which the escape velocity equals the speed of light. As a rule of thumb: The Schwarzschild radius of a black hole (in km) is approximately three times its mass in solar masses. So a 5-solar mass black hole has a Schwarzschild radius of about 5 × 3 = 15 km. The **event horizon** coincides with the Schwarzschild radius and is an imaginary boundary surrounding a collapsing star or black hole. Within the event horizon, no information of the events occurring there can be communicated to the outside.

Relativity

A full understanding of black holes and the phenomena associated with them requires knowledge of Einstein's theory of general relativity. Arguably Einstein's most famous works are his two theories relating time and space: special relativity and general relativity. Special relativity deals with the ultimate speed limit, the speed of light. And general relativity is a theory of gravity. General relativity gives a more complete description of gravity's effects than Newton's eighteenth-century description, and it can explain some anomalies that Newtonian mechanics cannot. Not that Newtonian mechanics is wrong; these days, it's just considered a special case of the more all-encompassing general relativity. Newtonian mechanics applies as long as the masses and velocities are not extraordinary.

Curved Space Ahead

Whereas Newton introduced the concept of gravitational force as a property of all matter possessing mass, Einstein proposed that matter does not merely attract matter, it warps the space around it. For Newton, the trajectory of an orbiting planet is curved because it is subject to the gravitational influence of, for example, the Sun. For Einstein, the planet's trajectory is curved because space itself has been curved by the presence of the massive Sun. This change represents a fundamental shift in the way we think about the universe. One major difference between Newton's and Einstein's theories of gravitation is that if mass distorts space, then massless photons of light (in addition to matter) should also feel the effect of gravity.

Like any good theory, Einstein's ideas explained questions that had gone unanswered, and made testable predictions. In particular, his theory explained some tiny but persistent peculiarities in the orbit of Mercury and successfully predicted that the Sun's mass was enough to bend light rays passing very close to it.

No Escape

One way to imagine space in Einstein's view is as a vast rubber sheet with a bowling ball sitting in a local depression. The mass of the bowling ball, representing a very massive object, distorts the sheet, which is a two-dimensional representation of space. In this model, a massive object distorts space itself.

Now, the sheet is two-dimensional. But space is three-dimensional. We can't picture that distortion applied to our three-dimensional universe, but we call its effect "gravity."

Imagine that instead of making a dimple in the sheet (as the bowling ball does), an object were to make an infinitely deep sinkhole. That region of space would be a black hole.

In the Neighborhood

We have said that no radiation can escape from within the Schwarzschild radius. But what of the region just outside it? It turns out that material near the black hole does produce observable radiation. Matter that strays close enough to the event horizon to be drawn into it does not remain intact, but is stretched and torn apart by enormous tidal forces. In the process, energy can be released in the form of x-rays. Bright x-ray sources are beacons that might point the way to black holes nearby.

Here's a Thought (Experiment)

No one could ever visit a black hole and live to tell about it. A spaceship, let alone a human being, would be torn to pieces by tidal forces as it approached the event horizon.

Faced with situations impractical or impossible to observe directly or to test physically, scientists typically construct *thought experiments*, methodical exercises of the imagination based on whatever data are available.

Star Words

A **thought experiment** is a systematic hypothetical or imaginary simulation of reality that is used as an alternative to actual experimentation when such experimentation is impractical or impossible.

Postcards from the Edge

So here's a thought experiment: Suppose it were possible to send an indestructible probe to the event horizon of a black hole.

Next, suppose we equipped the probe with a transmitter broadcasting electromagnetic radiation of a known frequency. As the probe neared the event horizon, we would begin to detect longer and longer wavelengths. This shifting in wavelength is known as a gravitational redshift, and it occurs because the photons emitted by our transmitter lose some energy in their escape from the strong gravitational field near the event horizon. The reduced energy would result in a reduction in frequency (and, therefore, a wavelength increase) of the broadcast signals—that is, a redshift.

The closer our probe came to the event horizon, the greater the redshift. At the event horizon itself, the broadcast wavelength would lengthen to infinity, each photon having used all of its energy in a vain attempt to climb over the event horizon.

Suppose we also equipped the probe with a large digital clock that ticked away the seconds and that was somehow also visible to us. (Remember, this is a *thought* experiment.) Through a phenomenon first described in Einstein's special relativity called *time dilation*, the clock would appear (from our perspective) to slow until it actually reached the event horizon, whereupon it—and time itself—would appear to slow to a crawl and stop. Eternity would seem to exist at the event horizon—the process of falling into the black hole would appear to take forever.

As the wavelength of the broadcast is stretched to infinite lengths, so the time between passing wave crests becomes infinitely long. Realize, however, that if you could somehow survive aboard the space probe and were observing from inside rather than from

Star Words

Time dilation is the apparent slowing of time (as perceived by an outside observer) as an object approaches the event horizon of a black hole, or moves at very high relative velocities, approaching the speed of light.

a distance, you would perceive no changes in the wavelength of electromagnetic radiation or in the passage of time. Relative to you, nothing strange would be happening. Moreover, as long as the physical survival of your craft in the enormous tidal forces of the black hole were not an issue, you would have no trouble passing beyond the event horizon. But to remote observers, you would have stalled out at the edge of eternity. It's all a matter of point of view.

Into the Abyss

What's inside a black hole, you ask? We have no idea. Not because we're not curious, mind you, but because we can literally get no information from beyond the event horizon.

Star Words

A **singularity** is the point of infinitely high density that is the remnant of a massive core collapse.

Theoreticians refer to the infinitely dense result of limitless collapse as a *singularity*.

We do not yet have a description of what happens to matter under the extreme conditions inside a black hole. Newton and Einstein moved our knowledge of the universe ahead in two great leaps. An understanding of these strange objects might be the next great leap forward in our comprehension of the universe.

The Latest Evidence

How can we actually *detect* a black hole? Can we see them at all?

Not directly, but we can certainly see their effects. Black holes are like some horror movie monsters. We might never see the monster itself, but evidence of its existence is everywhere: its footprints and claw marks on the trees, the muddy trail that leads back to the swamp. What are the latest muddy trails that lead us to black holes?

Wouldn't X-Rays Kill a Swan?

In our own galaxy, there is a bright source of x-rays in the constellation Cygnus (the Swan) known as Cygnus X-1, located at a distance of some 8,000 light-years. It is the binary companion of a B star, and the x-rays from it flare up and fade quickly, indicating that it is very small in radius. Remember that we mentioned the x-rays are often emitted in the neighborhood of a stellar-remnant black hole. In addition, the x-ray source has no visible radiation and a mass (inferred from the orbit of its companion) of about 10 solar masses. In this case circumstantial evidence might be enough to convict.

Black Holes in Our Own Backyard

The universe is a big place, so having a black hole in our own galaxy means that it is, relatively speaking, quite close. In fact, astronomers have detected a black hole and an old star that are orbiting one another only 5,000 light-years away from us. Poetically named GRO J1655-40, the black hole is an object referred to as a *microquasar*—a black hole that has the mass of a star and acts like a small version of the black holes that can have more than a million times the mass and are at the cores of active galaxies called *quasars*.

The black hole is apparently the remnant of a supernova explosion that the companion star survived, and the companion orbits the black hole every 2.6 days. What is perhaps most interesting about this pair of stars is that they are tearing through the Galaxy at high velocity—about four times the velocity that is normal for a star in this part of the Galaxy.

This image shows an artist's conception of a black hole known as a "microquasar," and its companion star.

(Image courtesy of ESA, NASA, and Felix Mirabel)

Now *That's* a Black Hole

The black hole in Cygnus is what is called a stellar remnant black hole. In theory, black holes can have much higher masses—masses far greater than a stellar core. These are called supermassive black holes and are found in the cores of galaxies. The *Hubble Space Telescope* image shown here is about as close as we have gotten to a supermassive black hole. Released on May 25, 1994, the image shows a whirlpool of hot (10,000 K) gas swirling at the center of an elliptical galaxy (see Chapter 17 for an explanation of elliptical galaxies) known as M87, 50 million light-years from us in the direction of the constellation Virgo.

Using the *Hubble's* Faint Object Spectrograph, astronomers Holland Ford and Richard Harms were able to measure how light from the gas is redshifted and blueshifted as one side of the 60-light-year-radius disk of gas spins toward us and the other away from us. The radius at which gas is spinning and the velocity of its rotation tell us how much matter must be within that radius. With the high resolution of the *Hubble Space Telescope* and radio telescopes like the Very Long Baseline Array (VLBA), astronomers can trace the rotation of gas to smaller and smaller distances from the center. Doing so, they find that even at very small radii, the gas is rotating at velocities that indicate something very massive still lies within that radius.

Recent observations have also suggested that, when galaxies collide, the black holes present at the centers of the two galaxies merge as well.

Black holes are almost certainly responsible for another observed galactic phenomenon. Radio astronomers have identified what they call "radio galaxies," which make themselves known by enormous *radio jets* that can extend from them for hundreds of thousands of light years. The sources of these jets are black holes. Now, when two galaxies collide and the black holes merge, the direction of the jet emitted can abruptly change, creating what is known as an X-type radio source. Thus, it appears that X marks the spot—at least for some black holes.

This image shows the radio jets of the source NGC 326 as imaged with the VLA. Inset is the HST image of the same galaxy.

(Image courtesy of NRAO/AUI, STScI)

The Least You Need to Know

♦ A neutron star is one possible remnant of a massive star that has exploded as a supernova.

♦ Black holes are the other possible remnant of collapsed massive star cores. If the core has a mass greater than three solar masses, the collapse cannot stop and a black hole is born.

♦ If you don't get too close to a black hole, its effects (in terms of gravity) are no different than a star of the same mass. They are *not* giant galactic "vacuum cleaners."

♦ According to general relativity, black holes, stars, and anything with mass distorts the space around it. This distortion can influence the path of both particles *and* light.

♦ Because light cannot escape from them, we can't see black holes, but we can indirectly observe evidence that their mass is present, or the jets that are believed to emerge from their vicinity.

Part 4

Way Out of This World

We turn from individual stars and stellar phenomena to some of the largest structures in the universe: the galaxies. We begin at home, with a chapter on our galaxy, the Milky Way, then move on to an explanation of the three major galaxy types—spiral, elliptical, and irregular—and how galaxies, themselves vast, are often grouped into even larger galaxy clusters, which, in turn, may be members of superclusters. Here is cosmic structure on the greatest scale.

The first two chapters in this part deal with normal galaxies, but there is another galaxy type, the active galaxy, which emits large amounts of energy. We explore such objects as Seyfert galaxies and radio galaxies, as well as one of the most spectacular cosmic dynamos of all, quasars. We also explore the nature of a phenomenon that appears to be occurring in distant galaxies: gamma ray bursters.

16

The Milky Way: Much More Than a Candy Bar

In This Chapter

- ◆ The Milky Way's birth, evolution, size, and motion
- ◆ Our place in the Milky Way
- ◆ The structure of the Milky Way
- ◆ A variety of variable stars
- ◆ The mystery of dark matter
- ◆ The galactic center: a black hole?

Ancient societies were well aware of the Milky Way, a fuzzy band that arcs across the sky, although they had no idea that their world was a tiny part of this very arc. Without the aid of a telescope, it isn't possible to see that this band actually consists of billions of individual stars. For this reason, ancient cultures described the Milky Way variously: as a bridge across the sky, as a river, as spilled corn meal from a sack dragged by a dog, or as the backbone of the heavens. So imagine Galileo's surprise when he first looked through his telescope at that white band of light: The fuzz was resolved into an enormous number of individual stars. Galileo must have been stunned.

Why were those stars all lying in an arc on the sky? It was several centuries before we determined our place in that grand arc.

With many of us living in light-polluted cities and suburbs, the Milky Way can be difficult or impossible to see. Under the best viewing conditions, it is a stunning sight, a majestic band of light sometimes extending high above the horizon.

When we look at the band, we see our own Galaxy from the inside. The Milky Way is our home, and, in this chapter we take a closer look at our place in the Milky Way, and how it might have come to be.

Where Is the Center and Where Are We?

Galileo and other astronomers soon realized that the stars around us were distributed in a distinct way, confined to a narrow band of the sky. In the late eighteenth century, William Herschel proposed the first model of our Galaxy, suggesting that it was disk shaped, and that the solar system (centered on the Sun) lay near the center of this disk. This model held some psychological comfort (Earth having been recently elbowed from the center of the universe by Copernicus and Galileo), but the model also had a problem. Herschel failed to account for what is now known as interstellar extinction. He had assumed that space was essentially transparent. But the disk of our Galaxy is "foggy" with gas and dust, and we can see only so far into the "fog."

As a result, we only *appear* to be at the center of the disk (as determined from star counts in various directions) not because we actually are, but because the dust in our Galaxy absorbs visible light so that we can only see out into the disk for a limited distance—the same limited distance in all directions. We clearly run out of stars in two directions when we look out of the "plane" of our Galaxy.

It was not until early in the twentieth century that we realized that we were not, in fact, at the center of our own Galaxy, but at its outer suburbs.

Home Sweet Galaxy

The universe is not evenly populated with stars and other objects. Just as there are large distances between individual stars, there are large distances between the collections of stars called galaxies. Galaxies (of which the Milky Way is one) are enormous collections of stars, gas, and dust.

radio continuum (408 MHz)

atomic hydrogen

radio continuum (2.5 GHz)

molecular hydrogen

infrared

mid-infrared

near infrared

optical

x-ray

gamma ray

NASA **Multiwavelength Milky Way**

This image shows the Milky Way Galaxy as imaged at a variety of wavelengths. The center of the Galaxy is in the center of each image. The optical image (labeled) is the most familiar view, but other wavelengths give us a unique view of our home Galaxy.

(Image courtesy of NASA)

Typical galaxies contain several hundred *billion* stars. There are about 100 times as many stars in our Galaxy as there are people on Earth. Astronomers conventionally refer to our own Galaxy with a capital "G" to distinguish it from all of the other galaxies in the universe.

A Thumbnail Sketch

Because we live within the Milky Way, we see it in profile. The Milky Way, viewed at this angle, is shaped rather like a flying saucer—that is, it resembles a disk that bulges toward its center and thins out at the periphery. The thickest part we call the *bulge*, and the thin part, the *disk*. (Astronomers aren't *always* in love with jargon!) You will see later how we figured this all out, but we know now that the center of our own Galaxy is in the *Galactic bulge*, in the direction of the constellation Sagittarius.

Our solar system is located in part of the thinned-out area, the so-called *Galactic disk*. Our location in the Galaxy explains what we see when we look at the Milky Way on a clear summer night in the country, far from city lights and smog. The wispy band of light arcing across the sky is our view *into* this disk. The band of light is the merged glow of the stars close enough to be seen optically—so

Star Words

The **galactic bulge** or **nuclear bulge** is a swelling at the center of our Galaxy. The bulge consists of old stars and extends out a few thousand light-years from the Galactic center.

Star Words

The **Galactic disk** is the thinnest part of our Galaxy. The disk surrounds the nuclear bulge, and contains a mixture of old and young stars, gas, and dust. It extends out some 50,000 light-years from the Galactic center, but is only about 1,000 light-years thick. It is the dust in the disk (a few hundred light-years thick) that creates the dark areas that run the length of the Milky Way and limits the view of our own Galaxy.

many that, to the unaided eye, they are not differentiated as separate points of light. A dark band of obscuration runs the length of the arc. This light is blocked by the presence of large amounts of dust in the disk.

When we look away from the arc in the sky, we don't see much of the Milky Way, because we are looking out of the disk. Looking into the disk is like looking at the horizon on Earth: There's lots of stuff there to block our view, including houses, trees, and cars. But look straight up into the sky, and you can see much farther—you might even see a plane flying overhead. When we look more or less perpendicular to the disk of the Galaxy, we can see much farther—this is the direction in which to look for other galaxies, for example.

It was while looking in this direction—up in the air, as it were—that astronomers discovered another component of the Galaxy, the *globular clusters*. Globular clusters are collections of several hundred thousand mostly older stars, held together by their mutual gravitational attraction. They are generally found well above and below the disk of the Galaxy. Reasoning that globular clusters should cluster around the gravitational center of the Galaxy, Harlow Shapley in the early twentieth century used the distances and positions of these collections of stars to determine where in the Galaxy we are located.

Along with these collections of stars in globular clusters are single stars.

This simple diagram of our view of the Milky Way shows the principal parts. Studies have shown that the Milky Way is an ordinary spiral galaxy.

Milky Way Galaxy
Seen edge-on

Compare and Contrast

Not even Dick Clark will live long enough to watch a single star evolve. So to overcome the limitation of the human life span, remember that we can chart the development of a certain mass star by observing many similar stars in various stages of development. The same is true for galaxies. The Milky Way (at our radius) takes about 225 million years to rotate just once! We can sometimes look for galaxies similar to our own in order to gain the necessary perspective and make generalizations about the Milky Way.

The Andromeda galaxy (Messier Object 31) is over 2 million light-years from us, and is the only object outside our Galaxy that is visible with the naked eye. Like the Milky Way, Andromeda is what we call a spiral galaxy, with distinct, bright, curved structures called spiral arms that reveal where the young, massive stars in the galaxy are forming. The universe contains many spiral galaxies that astronomers can observe to study aspects of the Milky Way that are hidden by our location in it.

Let's Take a Picture

Although Andromeda can be seen with only your eyes, or a modest amateur telescope, as a fuzzy patch of light, photographic equipment or an electronic CCD camera is required to make out the kind of detail in a galaxy that we are used to seeing in glossy astronomy magazine pictures. To see sweeping arms and dark bands of dust in other galaxies, we need to collect more light, either by using a larger aperture (a big telescope), or waiting longer (taking a sufficiently exposed conventional or electronic photograph).

Measuring the Milky Way

One way to gauge the size of the Milky Way would be to look at other, similar galaxies, such as Andromeda. If we know how far away such a galaxy is, and we can measure its angular size on the sky, we can calculate how big it is.

But determining the distance of Andromeda and other galaxies can be a serious challenge. Parallax as a distance indicator is out, because we know that the apparent angular shift is equal to 1 divided by the distance in parsecs. Stars farther than 100 pc are too far to use the parallax method, and those stars are still in our own galaxy. Half a microarcsecond resolution (0.0000005") would be required to see parallax of even the nearest galaxy, Andromeda. And no existing telescopes have resolution even approaching that. One very good distance indicator was discovered in 1908 by Henrietta Swan Leavitt, working at the time under the direction of Edward Pickering at the Harvard College Observatory.

Star Words

A **variable star** is a star that periodically changes in brightness. A **cataclysmic variable** is a star, such as a nova, that changes in brightness suddenly and dramatically, as a result of interaction with a binary companion star. An **intrinsic variable** changes brightness because of rapid changes in its diameter. **Pulsating variables** are intrinsic variables that vary in brightness in a fixed period or span of time.

Over many centuries of star gazing, astronomers had noted many stars whose luminosity was variable—sometimes brighter, sometimes fainter. These *variable stars* fall into one of two broad types: *cataclysmic variables* and *intrinsic variables*. We have already discussed one type of cataclysmic variable star. Novae are stars that periodically change in luminosity (rate of energy output) suddenly and dramatically when they accrete material from a binary companion.

Another type of variable star is an *intrinsic variable*, whose variability is not caused by interaction with a binary companion, but by factors internal to the star. The subset of intrinsic variable stars important in distance calculations are *pulsating variable* stars. Cepheid variable stars and RR Lyrae stars are both pulsating variable stars.

Astronomer's Notebook

The names of the class of variable stars, *RR Lyrae* and *Cepheid,* are derived from the names of the first stars of these types to be discovered. RR Lyra is a variable star (labeled RR) in the constellation Lyra; RR Lyrae is a genitive form of this name. The Cepheid class is named after Delta Cepheus, the fourth-brightest star, and a variable star in the constellation Cepheus. Cepheid variables have longer periods of brightness variation (as a class) and are more luminous than RR Lyrae stars. RR Lyrae stars have periods of less than a day. Cepheid periods range upward of 50 days.

The pulsating variables vary in luminosity because of regular changes in their diameter. Why does this relationship exist? Stars are a little like ringing bells. We know that, when struck, large bells vibrate more slowly, producing lower tones. Tiny bells vibrate rapidly, producing higher tones. In the same way, more massive stars vibrate slowly, and smaller stars vibrate more rapidly.

A pulsating variable star is in a late evolutionary stage and has become unstable, its radius first shrinking and its surface heating. Then its radius expands and its surface cools. These changes produce measurable variations in the star's luminosity (because luminosity depends on surface area, and the star is shrinking and expanding).

" " Close Encounter

Henrietta Swan Leavitt was one of a number of talented astronomers on the staff of the Harvard College Observatory. She was the first to propose, in 1908, that the period of a certain type of intrinsic variable star (a Cepheid variable) was directly proportional to its luminosity.

She observed a large number of variable stars in the Magellanic clouds (companions to our own galaxy), discovering over 1,700 of them herself. The advantage of studying these clouds is that by figuring their distance from Earth we are able to estimate the distance from Earth to those stars within them. What Leavitt noticed was that the brightest Cepheid variable stars in the Magellanic clouds always had the longest periods, and the faintest always had the shortest periods.

What her discovery meant was that astronomers could simply measure the period of a Cepheid variable star and its apparent brightness to derive its distance directly. What Leavitt accomplished was to greatly extend the astronomer's ruler from a few hundred light-years (using the parallax method) to tens of millions of light-years.

The two types of pulsating variables are named after the first known star in each group. The RR Lyrae stars all have the similar average luminosity of about 100 times that of the Sun. Cepheid luminosities range from 1,000 to 10,000 times the luminosity of the Sun. Because both pulsating variables can be recognized by their pulsation pattern and average luminosity, they make convenient markers for determining distance. Cepheid variable stars are intrinsically more luminous, so they are useful for measuring greater distances.

Armed with Leavitt's period-luminosity relation, American astronomer Harlow Shapley (1885–1972) used the period of RR Lyrae variable stars in almost 100 globular clusters in the Galaxy to determine their distances. With this distance information, Shapley could see that the globular clusters were all centered on a region in the direction of Sagittarius, about 25,000 light-years from Earth. He concluded from his study of globular clusters that the Galaxy was perhaps 30,000 parsecs (100,000 light-years) across—far larger than anyone had ever before imagined.

In an eerie twentieth-century replay of the days of Copernicus and Galileo, Shapley had demonstrated that the hub of the Galaxy, the Galactic center, was certainly not us. It lay fully 8,000 parsecs (25,000 light-years) from the Sun. Our star and our solar system are far from the center of it all.

Earlier in this chapter we defined various parts of the Milky Way: the nuclear bulge, the disk, the halo. Each of these regions has a characteristic size. Studies in this century have shown that the different parts of the Milky Way are also characterized by distinct stellar populations.

Astro Byte

Shapley made his discoveries about globular clusters in 1917, but it was not until Edwin Hubble observed Cepheid variables in the Andromeda galaxy, more than a decade later, that Andromeda and other so-called spiral nebulae were recognized as galaxies in their own right, separate from the Milky Way and very distant. Within a decade, we were told that not only were we not at the center of the Galaxy, but our Galaxy was revealed as only one of billions of galaxies. The universe was much more vast than we had ever imagined.

If you look at a true-color picture of the Milky Way or a similar galaxy, such as Andromeda, the bulge appears more yellow, and the disk appears more blue or white-blue. The color of a region tells us the average color of most of the stars that reside there. Globular clusters, the bulge, and the halo appear to contain mostly cooler (yellow) stars. Most of the young (hot) stars are gone, and new ones aren't taking their places in those locations.

The presence of large numbers of the youngest, hottest stars of type O and B makes the Galactic disk appear blue. Because O and B stars have such short lifetimes (1 to 10 million years), their presence in the disk tells us that they must be forming there "now" ("now" to an astronomer has a different meaning than it does to most people). Remember that the Milky Way (at our radius) rotates once every 225 million years. That means that tens of generations of massive O and B stars form and explode every time the disk of the Galaxy rotates. Cooler and smaller G, K, and M stars are also present in the disk, but the giant blue stars, far more luminous, outshine them, imparting to the entire disk region its characteristic blue-white color.

Why are the youngest stars in the disk and the oldest in the halo?

As you saw in Chapter 14, the Galactic disk is where the interstellar gas clouds reside—what we called giant molecular clouds. With raw materials plentiful, star production is very active here; therefore, young stars are abundant. In contrast, the halo region has very little nonstellar material, so that no new stars are being created there. In the Galactic bulge, there is an abundance of interstellar matter as well as old *and* new stars.

An image of the Milky Way from an edge-on perspective made by the Diffuse Infrared Background Experiment (DIRBE) on the Cosmic Background Explorer (COBE) *probe. The basic disk-bulge architecture of the Milky Way is apparent in this infrared image that highlights the location of low-mass stars and dust.*

(Image from NASA)

Milky Way Portrait

The different parts of the Milky Way are not static, but in constant motion. The disk rotates about the Galactic center, and at large radii, the rate of rotation does not trail off, but remains fairly constant. This rotation is in contrast to what we see in a planetary system, where planets rotate more and more slowly the farther they are from the center (think of Pluto). As you'll see in a moment, the constant rotation rate at large distances from the Galactic center betrays the presence of something we cannot see.

The stars in the halo move very differently, plunging through the Galactic disk in elliptical orbits that are randomly oriented. Their orbits are not confined to the Galactic disk, and in some ways seem unaware of it. All of these orbits are centered on what we call the Galactic center region and give clues as to how the Galaxy formed.

A Monster at the Center?

The central part of our Galaxy (first identified by Harlow Shapley) is quite literally invisible. It cannot be observed at optical ("visible") frequencies, but radio and infrared frequency observations have told us much about this hidden realm.

Some radio observations have identified a ring of molecular material orbiting the Galactic center at a distance of perhaps 8 or so parsecs (25 light-years). The rotational velocity of this ring tells us that there are several million solar masses of material located within its radius. Other radio observations (sensitive to hot gas) show that within this ring is a small amount of ionized material, nowhere near a few million solar masses.

Astro Byte

Our part of the Galactic disk orbits the Galactic center at over 136 miles per second (220 km/s). It takes about 225 million years for our region to complete one orbit around the Galactic center. Our solar system has orbited the Galactic center some 15 to 20 times since it formed. One quarter of a Galactic orbit ago, dinosaurs roamed Earth.

Recent infrared observations show that the Galactic center region contains a great many stars, closely packed. But they do not account for all of the required mass.

A few more clues: Radio astronomers have picked up strong radio emissions from a source in the Galactic center region that appears to be very small—perhaps as small as a solar system. Called Sagittarius A* (pronounced "A-star"), this source might give the location of a black hole of several million solar masses. This same source has been imaged with the new orbiting Chandra X-ray Observatory. The x-ray emissions from the Galactic center show the presence of hot gas and a jet, both of which are often associated with the presence of a black hole.

This x-ray image of the Galactic center taken with the Chandra X-ray Observatory shows the presence of two lobes of hot gas (labeled) perpendicular to the Galactic plane. These types of features are often associated with the presence of a black hole.

(Image courtesy of Chandra/NASA)

Arches Cluster

Quintuplet Cluster

1 light year

2 light years

Star Clusters Near the Center of the Galaxy   **HST · NICMOS**
PRC99-30 · STScI OPO · D. Figer (STScI) and NASA

Stars near the Galactic center are in orbit around a very small, very high-mass object. Is there a "monster" at the center? It appears increasingly likely that a black hole lurks at the Galactic center. These images show two clusters that are within 100 light-years of the center of the Galaxy.

(Image from D. Figer, STScI, and NASA)

The Birth of the Milky Way

The structure, composition, and motion of the Milky Way hold the keys to its origin. Though the theory of galaxy formation is far from complete, we have a fairly good picture of how our Galaxy might have formed.

Ten to fifteen billion years ago, an enormous cloud of gas began to collapse under its own gravity. The cloud, like most of the universe, would have been mostly hydrogen. It would have had a mass equal to that of all the stars and gas in the Milky Way, or several hundred billion solar masses.

The stars that formed first as the great cloud collapsed assumed randomly oriented elliptical orbits—with no preferred plane. Today, the oldest stars in our Galaxy—which are those in the Galactic halo and Galactic bulge—ring with the echoes of those early days. They are like the "asteroids" of our Galaxy. The Galactic halo is a vestige, a souvenir of the Galaxy's birth. And the globular clusters in the halo might have even formed prior to the cloud's collapse.

In a gravitational collapse process very similar to that which formed the solar system, the great cloud began rotating faster around a growing mass at the center of the Galaxy. The rotation and collapse caused the clouds of gas to flatten into the Galactic disk, leaving only those early stars in the halo.

Because the Galactic disk is the repository of raw materials, it is the region of new-star formation in the Galaxy. The Galactic halo is out of fuel and consists only of older (cool) redder stars. The orderly rotation of stars and gas in the Galactic disk stands in contrast to the randomly oriented orbits of stars in the halo and the Galactic bulge.

Dark Matters

"What you see is what you get," the popular saying goes. Such is not always the case in Galactic affairs.

Using Kepler's Third Law, we can calculate the mass of the Galaxy. This mass (expressed in solar masses) can be derived by dividing the cube of the orbit size (expressed in astronomical units, or A.U.) by the square of the orbital period (expressed in years). And Newton told us that at a given radius, all of the mass causing the rotation can be considered to be concentrated at a point at the center of rotation. For a system in "Keplerian rotation" (like a planetary system), we would expect the velocities of rotation to decrease as we looked farther and farther out, much as Jupiter orbits more slowly than, say, Mercury. Taking this approach, we find that the mass of the Milky Way *within* 15,000 parsecs of the Galactic center—that is, the *radius* of the visible Galaxy (diameter ~30,000 parsecs)—is 2×10^{11} solar masses.

We would expect the mass of the Galaxy to drop off precipitously when we run out of matter—at the visible outer edge of the Galaxy. But the puzzling fact is that *more* mass is contained *beyond* this boundary than within it!

Within a radius of 40,000 parsecs from the Galactic center, the mass of the Milky Way is calculated to be about 6×10^{11} solar masses. This means that there is as much, if not more, of the Milky Way unseen as there is seen. Spiral galaxies like the Milky Way have *mass-to-light ratios* ranging up to 10, even in the visible part of the disk. That is, luminous matter on very large scales accounts for only 10 percent of the matter that we "see" via gravity.

What is all this other stuff we cannot see directly?

Whatever its makeup, it apparently emits no radiation of any kind—no visible light, no x-rays, no gamma radiation. But it cannot hide completely. We see it because it has mass, and its mass affects the way in which the stars and gas of the Milky Way orbit.

Star Words

Mass-to-light ratio is the ratio of the total mass of a galaxy (detected by its gravitational effect) to the mass in luminous matter. If there were no "dark matter," the mass-to-light ratio would be 1.

Astronomers call the region containing this mass the dark halo. And the Milky Way is not alone in possessing such a region. Many, if not all, galaxies have the same signature in the rotation of their stars and gas. The dark halo presumably contains dark matter, a catchall term that is used to describe a variety of candidate objects. The truth is, we're not sure what dark matter is, but we know it's there because we clearly can see its gravitational effects.

The nature of dark matter is still one of the greatest mysteries of science. Some astronomers have suggested that difficult-to-detect low-mass stars (brown dwarfs or faint red dwarfs) might be responsible for the mass in this region—although recent *Hubble Space Telescope* observations have suggested an insufficient quantity of such objects to account for so much mass. It has recently been established that neutrinos do have nonzero mass, so their presence might contribute to dark matter. Yet others propose that dark matter consists of hitherto unknown subatomic particles, which pervade the universe. Nature has yet to reveal this mystery.

In the Arms of the Galaxy

We have referred to our own Galaxy several times in this chapter as a *spiral galaxy*. We will discuss galaxy classification in the next chapter, but what are these *spiral arms?* When observed from afar, they appear to be arcs of bright emission, curving out from the center of the Galaxy. There might be only two arms or more, and they might be loosely or tightly wound.

How do we know that the Milky Way also consists of spiral arms? Using the 21-cm hydrogen line discussed in the Chapter 14, the distribution of neutral hydrogen (by far the most abundant element in the Galactic disk or anywhere) has been plotted. Both position and velocity of the hydrogen clouds are required to make this plot, because it shows a dimension that is not on the sky, namely depth. These radio images confirm the spiral structure of the Milky Way.

Star Words

A **spiral galaxy** is characterized by a distinctive structure consisting of a thin disk surrounding a Galactic bulge. The disk is dominated by bright curved arcs of emission known as **spiral arms**.

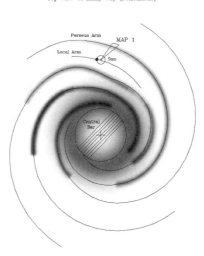

A schematic "top view" of the Milky Way Galaxy, showing the location of spiral arms as determined with the use of radio telescopes. Several surveys are now imaging the neutral hydrogen (HI, or "H-one") in the plane of the Galaxy.

(Image courtesy of Canadian Galactic Plane Survey)

These radio studies also confirm that we are far from the Galactic center, located unostentatiously on a cusp between two spiral arms. Spiral arms themselves are not that hard to account for, but their longevity is. In the next chapter, we will discuss how spiral arms might arise, and how they could possibly persist.

A detail of the plane of our Galaxy as seen in neutral hydrogen (HI) comes from one of the projects mapping our Galaxy, the Canadian Galactic Plane Survey (CGPS). This image is of a star-forming region in the direction of the constellation Cygnus.

(Image courtesy of J. English, A. R. Taylor/CGPS)

The Least You Need to Know

- Our home Galaxy is the Milky Way. We see it from within as an arc of fuzzy emission across the sky.

- The main components of the Galaxy are the disk, the stellar halo (containing the globular clusters), the nuclear bulge, and the dark halo.

- Our solar system is in the Galactic disk of the Milky Way, 25,000 light-years from the Galactic center.

- There is a large amount of evidence that the center of our Galaxy harbors a massive black hole.

- We can only account for a fraction of our Galaxy's mass with normal matter; most of the Galaxy's mass must be contained in a halo of material, consisting of "dark matter."

- Astronomers have built a map of the gas in the disk of our Galaxy using radio telescopes.

Chapter 17

A Galaxy of Galaxies

In This Chapter

- ◆ Hubble's classification of galaxy types: spiral, elliptical, and irregular
- ◆ Spiral density waves
- ◆ Determining the distance of galaxies
- ◆ Galactic clusters and superclusters
- ◆ Calculating the mass of galaxies and galactic clusters
- ◆ Receding galaxies, expanding universe
- ◆ Hubble's Law

Viewed with the naked eye, some "stars" appear fuzzier than others. After the invention of the telescope, it became obvious why this is the case. Through a telescope eyepiece, some apparent "stars" are resolved into the disks of planets, some into regions where stars are forming, and others into collections of old stars. There was great disagreement about one class of objects identified in the eighteenth and nineteenth centuries and called *spiral nebulae*. These were not stars, and yet there was no way to figure out how *big* they were without knowing how *far away* they were.

Before Edwin Hubble extended our conception of the size of the universe in the 1920s, these objects were classified as various nebulae and were generally thought to lie within the Milky Way, which, in turn, was believed to be synonymous with the universe. We now know that our Galaxy is but one of many, and that the spiral nebulae are entire other galaxies, some smaller and some bigger than our own, all containing hundreds of billions of stars. And we see galaxies no matter where we look in the sky. When we look deep enough, they are there. Some are so close that they are part of our "Local Group"; others are so distant that their light has barely had time to reach us since the universe began.

In this chapter and the next, we talk about galaxies other than our own. Astronomers have two major goals in the exploration of other galaxies. The first is to understand the galaxies themselves and how they have evolved with time. The second is to use other galaxies to tell us more about the Milky Way, our galactic home.

Sorting Out the Galaxies

A big part of the ongoing excitement of astronomy is the invigorating sensation of awe it can stir. News from other planets in the solar system is interesting, but it has a certain familiarity. However strange, a planet like Venus is about the size of Earth, and has a surface and an atmosphere. We can deal with that.

But galaxies are different. Each one of them contains several hundred billion stars. Does that number make you go blank? Remember this: An average galaxy contains about 100 times as many stars as there are people on our planet.

And there are a lot of galaxies out there.

Fortunately, galaxies are not all completely different, but fall into broad groups. They can be classified, just like trees or beetles or clams. But they couldn't be classified until there were photographs of large numbers of them, starting in the mid- to late nineteenth century.

The great American astronomer Edwin Hubble (1889–1953) came to the Mount Wilson Observatory in California in 1922 to use its new 100-inch (2.5-meter) telescope to study nebulae. A fundamental question remained unanswered back then. Were these spiral nebulae as big as our Galaxy and incredibly distant (as first suggested by Immanuel Kant's *Universal Natural History* and *Theory of the Heavens* in the eighteenth century)? Or were they more mundane, relatively nearby objects? The huge surface area and high resolution of the 100-inch telescope allowed astronomers to see that these nebulae (like our own Milky Way) resolved into individual, faint stars.

After intrinsic variable stars were identified in some of these "nebulae," the relationship between period and luminosity (see Chapter 16) could be used to determine their distance. And Hubble's distances, calculated from Cepheid variable stars, prompted him to conclude in 1924 that many of the nebulae were not part of the Milky Way at all, but were galaxies in their own right. Andromeda, for example, was thought to be almost a million light-years away, *far* outside the limits of our own galaxy (this distance has since been revised upward). Shortly after making his mind-blowing discovery, Hubble—not easily overwhelmed—set about classifying the galaxies he saw.

Hubble's classifications (like many first impressions) are based on appearance and fall into three broad categories: spiral, elliptical, and irregular.

Spirals: Catch a Density Wave

In the preceding chapter, we took a look at one example of a spiral galaxy, the Milky Way. Hubble labeled all spiral galaxies with the letter "S" and added an a, b, or c, depending on the size of the galactic bulge. Sa galaxies have large bulges, Sb medium-sized bulges, and Sc the smallest bulges. Very clearly defined, tightly wrapped spiral arms are also associated with Sa galaxies, whereas Sb galaxies exhibit more diffuse arms, and Sc galaxies have even more loosely wrapped, less clearly defined spiral arms.

Another subtype of the spiral galaxy is the *barred-spiral galaxy*, which exhibits a linear bar of stars running through the galactic disk and bulge out to some radius. In these barred-spiral galaxies, the spiral arm structure begins at the ends of the bar. This spiral subgroup, designated SB, also includes the a, b, and c classifications based on the size of the galactic bulge and the winding of the arms. The Milky Way Galaxy appears to have a bar structure near its center.

Star Words

A **barred-spiral galaxy** is a spiral galaxy that has a linear feature, or bar of stars, running through the galaxy's center. The bar lies in the plane of the spiral galaxy's disk, and the spiral arms start at the end of the bar.

Barred Spiral Galaxy NGC 1365
NASA and M. Carollo (Columbia University) • STScI-PRC99-34a
HST • WFPC2 • NICMOS

A region in NGC 1365, a barred-spiral galaxy located in a cluster of galaxies called Fornax. The central image was photographed through a ground-based telescope. The images to the left and right show the central bulge as seen at optical and infrared wavelengths with the Hubble Space Telescope.

(Image from NASA and M. Carollo)

Ellipticals: Stellar Footballs

Elliptical galaxies present a strikingly different appearance from the spirals. When viewed through a telescope, they look a bit dull compared to their flashy spiral cousins. There are no spiral arms, nor any discernable bulge or disk structure. Typically, these galaxies appear as nothing more than round or football-shaped collections of stars, with the most intense light concentrated toward the center and becoming fainter and wispier toward the edges.

Star Words

An **elliptical galaxy** is one that has no discernible disk or bulge, and looks like an oval or circle of stars in the sky. The true shapes of ellipticals vary from elongated ("footballs") to spherical ("baseballs") to flattened ("hamburger buns"). Elliptical galaxies consist mostly of old stars, and appear to have little gas left.

Of course, the orientation of an elliptical galaxy will influence its shape in the sky—that is, its apparent shape might not be its true shape. A football viewed from the side has a sort of oval shape, but, viewed end-on, looks like a circle. Regardless of true shapes, Hubble differentiated within this classification by apparent shape. E0 ("E-zero") galaxies are almost circular, and E7 galaxies are very elongated (or elliptical). The rest—E1 through E6—range between these two extremes. Astronomers can use observations of the motions of stars in elliptical galaxies, combined with computer modeling, to determine the true shapes of elliptical galaxies.

Hickson Compact Group (HCG) 87 consists of four galaxies. The edge-on spiral at the bottom of the image and the elliptical to the right are both known to have "active nuclei," most likely related to the presence of a black hole.

(Image from AURA/STScI/NASA)

There is no such thing as a typical size for an elliptical galaxy. Their diameters range from a thousand parsecs (these dwarf ellipticals are much smaller than the Milky Way) to a few hundred thousand parsecs (giant ellipticals). The giant ellipticals, often located in the center of galaxy clusters, are many times larger than our galaxy.

Like any classification scheme, Hubble's has its share of duck-billed platypuses, objects that don't quite fit in. Some elliptical-shaped galaxies exhibit more structure than others, showing evidence of a disk and a galactic bulge. They still lack spiral arms and, like the other ellipticals, they don't contain star-forming gas. This type of galaxy (since it does have a disk and a bulge) is designated S0 (pronounced "S-zero"). Some of these galaxies even contain a bar, and are designated SB0.

Are These Reduced? They're All Marked "Irregular"

Finally comes the miscellaneous category of *irregular galaxies*, which lack any regular structure. Galaxies in this class often look as if they are coming apart at the seams or are just plain messy. Unlike the ellipticals, irregulars are rich in interstellar material and are often sites of active star formation.

Two of the most famous irregular galaxies are very close to us: the Large and Small Magellanic Clouds. Usually called the LMC and the SMC by acronym-happy astronomers, they are visible from the southern hemisphere. These Milky Way companions (10 times closer to us than the closest spiral galaxy) interact with each other and our Galaxy via gravity, and both contain active star-formation regions. The Small Magellanic Cloud is stretched into an elongated shape due to the tidal forces exerted on it by our galaxy.

Star Words

Irregular galaxies lack obvious structure, but tend to have lots of raw materials for the creation of new stars.

Close Encounter

Those of us confined to Earth's northern hemisphere don't get to see the most spectacular of the irregular galaxies, the Small and Large Magellanic Clouds. Named for the sixteenth-century explorer Ferdinand Magellan, whose men brought word of them to Europe at the conclusion of their global voyage, the Magellanic Clouds are rich in hydrogen gas. The clouds, like moons, are believed to orbit the much larger Milky Way. Like any object with mass in the universe, galaxies feel the irresistible tug of gravity and are pulled into groupings by its effects.

Galactic Embrace

In the constellation Canes Venatici is Messier object 51 (M51), the Whirlpool galaxy. Its distinct shape resembles an overhead view of a hurricane or, less dramatically, a pinwheel. Bright curved arms grow out of its hublike galactic bulge and wrap partway around the galactic disk. It is easy to see the spiral pattern in the Whirlpool galaxy because it faces us at an angle that gives us a spectacular overhead view rather than a side view.

The spiral Whirlpool galaxy M51 as seen with the Very Large Array (VLA) and the Effelsberg 100-m at 6 cm. The short lines in the image to the right show the orientations of the magnetic fields embedded in material in the spiral arms.

(Image courtesy of NRAO/AUI)

It is the existence of these spiral arms that give such galaxies as the Whirlpool (M51), Andromeda (M31), and our own Milky Way the name spiral galaxies. But what exactly are spiral arms?

Their existence does indeed present a puzzle. We know that the stars and other matter in the galactic disk orbit differentially—faster toward the center, somewhat slower toward the periphery. This rotational pattern soon stretches any large clump of stars or gas in the disk of a spiral galaxy into a spiral structure. It does this so quickly that if differential rotation were to account for spiral arms, the arms would rapidly get totally wrapped up around a galaxy's bulge and disappear. Somehow, the Milky Way and other spiral galaxies retain their spiral arm structure for long periods of time, long enough that they are plentiful in the universe.

Most astronomers are now convinced that the arms of spiral galaxies are the result of compressions, called *spiral density waves*, moving through a galaxy's disk. These ripples in the pond of a galaxy's disk

Star Words

Spiral density waves are waves of compression that move around the disk of a spiral galaxy. These waves are thought to trigger clouds of gas into collapse, thus forming hot, young stars. It is mostly the ionized gas around these young, massive stars that we observe as spiral arms. Close encounters between galaxies can produce these spiral density waves.

move around the galaxy center, compressing clouds of gas until they collapse and form stars. It is mostly the ionized gas surrounding hot, young stars (formed by the passage of a density wave) that we see as spiral arms in the optical part of the spectrum.

The theory of spiral density waves neatly resolves the problem of the effect of differential rotation. As you might recall from our discussion of waves in Chapter 4, a wave is a disturbance that moves through matter, a ripple in a pond. Thus, spiral density waves move *through* the matter of a galactic disk and are not caught up in differential rotation. The stars that they form might then get stretched out by this differential rotation, but the waves keep on moving.

How to "Weigh" a Galaxy

In the same way that we cannot directly measure the temperature of stars, we cannot directly measure their mass. We measure the mass of the Sun, for instance, by using Kepler's Third Law. If we know the distance to the Sun and the period of Earth's orbit, we can calculate the Sun's mass. Galaxy masses are calculated in the same way, only the objects orbiting are stars and gas instead of planets.

A Big Job

Plotting the *rotation curve* of an individual spiral galaxy—the velocity of disk material versus the distance of that material from the galactic center—can yield the mass of the galaxy that lies within that radius. This method gives a good estimate as long as most of a galaxy's mass is contained near its center. In the solar system, for example, the outer planets rotate much more slowly than the inner planets, in accordance with Kepler's Third Law, because the mass of the solar system (99.9 percent of it, anyway) is contained in the Sun.

But astronomers soon noticed a problem with galaxies. There are objects in the outer reaches of spiral galaxies (clouds of gas—in particular, clouds of neutral hydrogen called HI clouds) that orbit in a way that indicates that they "see" more mass out there than we do here at the radius of the Sun. They are orbiting faster than they should. And using the 21-cm radio line of hydrogen to see the HI, astronomers have traced the rotation curve of many galaxies far beyond the outermost stars. These curves seem to indicate that

> **Astronomer's Notebook**
>
> Most spiral galaxies contain from 10^{11} to 10^{12} solar masses of matter, much of it located far beyond the visible radius of the galaxy. Large elliptical galaxies contain about the same mass (but can be even more massive), whereas dwarf ellipticals and irregular galaxies typically contain between 10^6 and 10^7 solar masses of material.

there is more matter at large radii in these galaxies. Whatever has this mass, though, is something that we can't see, because it is not "shining" at any wavelength. The rotation of galaxies indicates the presence of *dark matter.*

On the scale of clusters of galaxies, the mass-to-light ratio, which is about 10 in galaxies, can be 100 or more. That is, luminous matter on very large scales accounts for only 1 percent of the matter that we "see" via gravity.

"It's Dark Out Here"

Because dark matter accounts for a great deal of the mass of a galaxy—up to 10 times more than the mass of visible matter—the shocking conclusion must be that 90 percent of the universe is dark matter, utterly invisible in the most profound sense of the word. For it neither produces nor reflects any electromagnetic radiation of any sort at any wavelength.

You can think of it as an embarrassing truth, or as an exciting unanswered question. The fact remains: We're not quite sure what 90 percent of the matter in the universe is made of.

Let's Get Organized

Human beings are inveterate pattern makers. Men and women have looked at the sky for centuries and have superimposed upon the stars patterns from their own imaginations. Although the mythological heritage of constellations is still with us, scientists long ago realized that the true connections among the planets and among the stars are matters of mass and gravitational force, not likenesses of mythological beings.

Are there gravitationally determined patterns to the distribution of galaxies throughout the universe?

Measuring Very Great Distances

In Chapter 13, you saw that the distance from Earth to the planets can be measured accurately by radar ranging, but that measuring the distance to farther objects, namely the nearer stars, requires measuring stellar parallax. But beyond about 100 parsecs, parallax doesn't work well, because the apparent angular displacement becomes smaller than the angular resolution of our best telescopes.

Gas velocities and a model of the rotation of the Milky Way can be used to measure distances within our own Galaxy (out to about 30,000 light-years). Beyond this, and out to 10 to 20 million parsecs (30 to 60 million light-years), variable stars (the RR Lyrae and Cepheid variables discussed in Chapter 16) can be identified and observed in order to determine distance.

The trouble is Cepheid variable stars farther than 15 million parsecs are difficult to resolve and, for most telescopes, too faint to be detected. And many galaxies are well beyond 15 million parsecs away.

Two methods have been used to estimate intergalactic distances greater than 15 million parsecs. One tool is called the Tully-Fisher relation, which uses an observed relationship between the rotational velocity of a spiral galaxy and its luminosity. By measuring how fast a galaxy rotates (astronomers use the 21-cm hydrogen line described in Chapter 14), one can calculate its luminosity with remarkable accuracy. Once you know the luminosity of a galaxy, you can measure its apparent brightness and easily calculate its distance out to several hundred million parsecs.

The other tool is more general and involves identifying various objects whose luminosity is known. Such objects are referred to as *standard candles*. If we truly know the brightness of a source, we can measure its apparent brightness and determine how distant it is. A 100-watt light bulb is an example of a standard candle. It will look fainter and fainter as it recedes from us (getting dimmer as distance squared), and in fact, we can determine how far it is by measuring its apparent brightness versus its known luminosity.

One very interesting standard candle is a Type Ia supernova. It is interesting because its peak luminosity is very regular—and enormous (10 billion or 10^{10} solar luminosities!). We have discussed the Type II super-novae, the core collapse of a massive star. Type Ia supernovae occur when a white dwarf accumulates enough material from its binary red giant companion to exceed 1.4 solar masses. When this happens, the white dwarf begins to collapse, and the core ignites in a violent burst of fusion. The energies are sufficient to blow the star apart. These events are so luminous that we can see them *billions* of light-years away.

> **Star Words**
>
> A **standard candle** is any object whose luminosity is well-known. Its measured brightness can then be used to determine how far away the object is. The brightest standard candles can be seen from the greatest distances.

The Local Group and Other Galaxy Clusters

Armed with an ability to measure very great distances, we can begin looking at the relationships among galaxies.

Within 1 million parsecs (3 million light-years) of the Milky Way lie about 20 galaxies, the most prominent of which is Andromeda (M31). This galactic grouping, called the *Local Group*, is bound together by gravitational forces.

Star Words

The **Local Group** is a **galaxy cluster**, a gravitationally bound group of galaxies, which includes the Milky Way, Andromeda, and some 18 other galaxies.

The generic name for our Local Group is a *galaxy cluster*, of which thousands have been identified. Some clusters contain fewer than the 20 or so galaxies of the Local Group, and some contain many more. The Virgo Cluster, an example of a rich cluster, is about 15 million parsecs from the Milky Way and contains thousands of galaxies, all bound by their mutual gravitational attraction. Giant elliptical galaxies are often found at the centers of rich clusters where there are very few spirals.

From the velocities and positions of galaxies in clusters, one thing is very clear. We cannot directly observe at least 90 (perhaps as much as 99) percent of the mass that must be there. Galaxy clusters, like the outer reaches of spiral galaxies, must be mostly dark matter.

The galaxy cluster called Abell 2218 is about two billion light-years from Earth. When these photons left the cluster, life on Earth was very simple indeed. The effect of the "unseen" mass in this cluster can be seen as the arcs of light that are distorted light from background sources. This distortion, much like viewing a room through a wine glass, is sometimes referred to as "lensing."

(Image from NASA, A. Fruchter, and the ERO team)

Galaxy Cluster Abell 2218   **HST • WFPC2**
NASA, A. Fruchter and the ERO Team (STScI, ST-ECF) • STScI-PRC00-08

Superclusters

Galaxy clusters themselves are grouped together into what are called *superclusters*. The Local Supercluster, which includes the Local Group, the Virgo Cluster, and other galaxy clusters, encompasses some 100 million parsecs.

3.6Z light years

On the very largest scales that we can measure (hundreds of millions of light-years across), the universe has a bubbly appearance, with superclusters concentrated on the edges of large empty regions or voids. In the final chapters of this book we will explore the possible explanations for this large-scale structure.

Star Words

A **supercluster** is a group of galaxy clusters. The Local Supercluster contains some 10^{15} solar masses.

Where Does It All Go?

The galaxies within a cluster don't move in an orderly manner. Like the stars in the galactic halo that envelops a spiral galaxy, the galaxies appear to orbit the cluster center in randomly oriented trajectories. But on larger scales, beyond the confines of a single cluster, there does appear to be orderly motion. This motion is the echo of a cataclysmic event: the event that brought the universe into being.

Hubble's Law and Hubble's Constant

It has been known since the early twentieth century that every (sufficiently distant) spiral galaxy observed exhibits a redshifted spectrum; that is, all of these galaxies are moving away from us. Recall from Chapter 13 the Doppler effect: Wavelengths grow longer (redshift) as an object recedes from the viewer. The conclusion is inescapable: All galaxies partake in a *universal recession*. This is the apparent general movement of all galaxies away from us. This observation doesn't mean that we are at the center of the expansion. Any observer located anywhere in the universe sees the same redshift.

In 1931, Edwin Hubble and Milton Humason first plotted the distance of a given galaxy against the velocity at which it receded. The resulting plot was dramatic. The rate at which a galaxy is observed to recede is directly proportional to its distance from us; that is, the farther away a galaxy is from us, the faster it travels away from us. This relationship is called Hubble's Law. This law relates the velocity of galactic recession to its distance from us. Simply stated, Hubble's Law says that the recessional velocity is directly proportional to the distance.

Picture a bunch of dots on the surface of a toy balloon. Our galaxy is one of the dots, and all of the other galaxies in the universe are the other dots. As you inflate the balloon, the surface of the balloon stretches, and, from the point of view of any dot (galaxy), all the other dots (galaxies) are moving away. The farther away the dot, the more balloon there is to stretch, so the faster the dot will appear to recede.

One benefit of Hubble's Law is that it can be used to extend our cosmic distance scale to extraordinary distances. Although the Tully-Fisher technique will get us out to about 200 million parsecs, and Type Ia supernovae to a few billion parsecs, Hubble's Law goes farther. In fact, Hubble's Law gives us the distance to any galaxy for which we can measure a spectrum (in order to get the Doppler shift). With the velocity from that spectrum, and the correct value of the *Hubble constant*, we arrive quite simply at the distance.

But there is a twist. We have assumed that the universal expansion happens at a constant rate. Is this a good assumption?

Star Words

Hubble's Law turns on **Hubble's constant** (H_0), the constant of proportionality between the velocity of recession and the distance from us. The value of H_0 is expressed in kilometers per second per megaparsec (a megaparsec [Mpc] is 1 million parsecs) and is found to be about 65 km/s/Mpc, meaning that the universe has an age of about 14 billion years.

Recent observations suggest that the universe has not always been expanding at the same rate. Two magnificent forces are in opposition. There is the expansion of the universe (as we will see, set into play by the Big Bang), and there is the force of gravity, pulling every two particles with mass together, diminished by distance squared. If there were nothing pushing things apart, the universe would, due to gravity, collapse to a point. This would be a problem, and Einstein recognized it as such. When he was working on general relativity, he needed to add a term—the cosmological constant—to his equations to balance the force of gravity, to keep the universe "puffed up." This term was needed to keep the universe static, as it was then thought to be.

But then Edwin Hubble came along and showed that the universe was hardly static; in fact, everything was rushing away from everything else, and although gravity might eventually win out, the universe wasn't going to collapse to a point any time soon. Einstein later called the introduction of the cosmological constant into his equations his "greatest blunder."

As it turns out, however, Einstein's "blunder" might not have been a mistake at all. The cosmological constant might actually be needed. Recent observations made by scientists of very distant ("high-redshift," in astronomer parlance) Type Ia supernovae show something very surprising. These "standard candles" can be seen out to 7 billion light-years, and the distances to the galaxies that they are in show that the universe appears to have been expanding more slowly in the past than now. So not only is everything rushing away from everything else, but it's rushing away more quickly these days. In other words, the expansion of the universe is accelerating.

What is causing it to accelerate? Well, it might be the "vacuum energy" that Einstein thought he needed to support the universe from collapsing on itself. Interestingly, this force increases with distance (as opposed to gravity, which weakens with distance). This new energy required to power an accelerating expansion has been dubbed "dark energy," implying that, as with "dark matter," we know it is there, but we are not sure what exactly it is.

The Big Picture

Expansion implies a beginning in time. And we will explore where and how the expansion might have begun in Chapter 19. In Chapter 20, we'll explore the details of the expansion itself. But before we move on to some of these big questions, we turn our attention in the following chapter to some of the most energetic and unusual members of the galactic family: the quasars, black holes (again), and galactic jets, all of which keep the universe quite active, thank you very much.

The Least You Need to Know

- Galaxies were first classified into three broad types by Edwin Hubble: spiral, elliptical, and irregular.

- The majority of the mass (90 percent) of most galaxies and clusters (99 percent) is made of material we cannot observe, dark matter. We see it only by its gravitational effect.

- Distances on intergalactic scales are measured by observation of Cepheid variable stars, the Tully-Fisher relation, standard candles such as Type I supernovae, and Hubble's Law.

- Many galaxies are grouped in galactic clusters, which, in turn, are grouped into superclusters. Superclusters are found together on the edges of huge voids.

- Edwin Hubble first observed that galaxies recede from us at a rate proportional to their distance from Earth. The constant of proportionality, called the Hubble constant, tells us the age of the universe.

- The expansion of the universe appears to be accelerating, powered by an as yet unexplained "dark energy."

Chapter 18

Moving Out of Town

In This Chapter

- Distant galaxies, strange galaxies
- Active galaxies: Seyfert and radio galaxies
- What drives an active galaxy?
- Galactic jets
- What's in a quasar?

Remember Grote Reber? He's the man who, back in 1936, built a radio telescope in his backyard, and by the 1940s had discovered the three brightest radio sources in the sky. He didn't know it at the time, but two of them—the Galactic center, Sagittarius A; and the shrapnel of a supernova explosion, Cassiopeia A—are sources in our own galaxy. But the third radio source, called Cygnus A, turned out to be much, much farther away and far more strange.

In 1951, the pioneering German-American radio astronomers Walter Baade and Rudolph Minkowski located a dim optical source at the position of Cygnus A and, from its spectrum, measured a redshift (or its recessional velocity) of some 12,400 miles per second (almost 20,000 km/s). Whatever it was, Cygnus A was moving away from us—fast! For a while, astronomers tried to figure out how an object in our galaxy could be moving so fast. It turns out it wasn't in our Galaxy at all, but very far away. Later, astronomers would discover that this was something never before seen: a distant inferno churning out the energy of a hundred normal galaxies.

In this chapter, we will examine objects like Cygnus A and the other monsters that hide deep in the hearts of distant galaxies. They are some of the most energetic and bizarre objects in the universe.

A Long Time Ago in a Galaxy Far, Far Away ...

Because of their great distances—hundreds of millions of parsecs away—the farthest "normal" galaxies are very faint. (We'll talk about what *normal* means in just a moment.) At such extreme distances, it becomes difficult even to see such galaxies, much less study their shapes or how they are distributed in space. But as far as we *can* see, there is little difference between normal distant galaxies and normal galaxies closer to home.

The operative word here is *normal*. Out in the farthest reaches of space, we see some objects that are not normal—at least they are not what we're accustomed to seeing in our cosmic neighborhood.

What does this tell us? Remember that the universe has a speed limit, the speed of light, and that information can travel no faster than this speed. Thus the farther away a certain star or galaxy might be, the longer it has taken its light (and the information it contains) to reach us.

Thus not only are we seeing far into space, we are in a very real sense seeing *back into time*. The strange objects that lurk in the distant universe existed in its earliest times. To study these objects—*quasars*—is to peer into the origin of galaxies, including perhaps the origin of our own Milky Way.

Quasars: Looks Can Be Deceiving

After seeing the spectacular optical images of emission nebulae (see Chapter 14) and galaxies such as Andromeda (see Chapter 16), a *quasar* can make a disappointing first impression. Although the optical counterparts of the brightest of them can be seen with an amateur telescope on a dark night, they are undistinguished sources of light. In fact, they look so much like stars that astronomers at first thought they were simply peculiar stars (they were dubbed "quasi-stellar objects").

But keep in mind how incredibly distant these objects are. The closest quasars are some 700 million light-years away. Though their apparent brightness might be small, their luminosities (the amounts of energy that they put out each second) are astounding.

Quasars appear optically faint for two reasons: (1) much of their energy is emitted in the nonvisible part of the spectrum, and (2) quasars are very distant objects.

The great distance to quasars became truly apparent in the 1960s. The first quasars were discovered at radio frequencies, and optical searches at these locations showed objects that looked like stars. But the spectra of these "stars" told a different story. In the early 1960s, the astronomer Maarten Schmidt made a stunning proposal. The four bright spectral lines that distinguish hydrogen from the rest of the elements in the universe were seen to be shifting to longer wavelengths— much longer wavelengths. These *redshifted* lines, along with Hubble's Law, indicated that quasars were very distant—billions of light-years away, in fact.

Star Words

Quasar is short for "quasi-stellar radio source." The first quasars were detected at radio frequencies, though most quasars don't emit large amounts of radio energy. Quasars are bright, distant, tiny objects which produce the luminosity of 100 to 1,000 galaxies within a region the size of a solar system.

This image from the Very Large Array shows the detailed structure of Cygnus A, located in the constellation Cygnus, the swan. One of the brightest radio sources in the sky, Cygnus A is one of a large number of radio galaxies that has been discovered, with "jets" that arise from the region around a central black hole.

(Image from NRAO)

Quasar 3C 273 (the 3C stands for the Third Cambridge Catalogue) has its spectral lines redshifted in velocity by 16 percent of the speed of light. As we know, light from a source is redshifted when the object is moving away from us. In the context of Hubble's Law, such dramatic redshifts mean that quasars are receding at tremendous speeds and are very far away. The quasar 3C 273 travels at some 30,000 miles per second (48,000 km/s) and is some 2 billion light-years (640 million parsecs) distant from us.

It is an awe-inspiring thing to contemplate an object so powerful and so distant. Quasars are distant and bright, but also small. How do we know they are small? Many quasars flicker—varying in brightness, rapidly—on scales of days. We know that light must be able to travel across the size of an object for us to see it vary in brightness. Why? Because for a region to appear to brighten, you must detect photons from the far side as well as from the near side of the object that is brightening. If an object is one light-year across, the "brighter" photons from the far side of the source will not reach you until a year after the photons on the near side. So this source could only flicker on scales of a year.

> ### Astronomer's Notebook
>
> Quasars, among the most luminous objects in the universe, have luminosities in the range of 10^{38} watts to 10^{42} watts. These numbers average out to the equivalent of 1,000 Milky Way Galaxies.

Any object that flickers on scales of mere days must be small—light-*days* across, to be exact. This characteristic flickering reveals that the source of energy is perhaps the size of our solar system—presumably a gaseous accretion disk—a collection of interstellar material—caught in the gravitational field of a supermassive black hole and spiraling toward it.

If a supermassive black hole is the source of a quasar's power, then about 10 Sunlike stars per year falling into the black hole could produce its enormous luminosity.

Quasars and the Evolution of Galaxies

Quasars might be more than strange, distant powerhouses of the early universe. In fact, they might be part of the family tree of every galaxy, including our own. Perhaps all galaxies (as they form) start out as quasars, which become less luminous, less energetic, as the early fuel supply of the galaxy is exhausted. Certainly quasars cannot burn fuel forever at the prodigious rates that they do.

But what could fuel a quasar? A generation of stars that forms near the center of the galaxy early in its life could, through the natural mass loss that occurs as stars age, provide the needed mass. Another possibility is a galactic train wreck. If two galaxies collide, one could provide fuel for the other's dormant black hole. As the fuel arrives, the black hole once again lights up. Bingo! A quasar.

Cartwheel Galaxy
PR95-02 · ST ScI OPO · January 1995 · K. Borne (ST ScI), NASA

HST · WFPC2
12/23/94 zgl

When galaxies collide, star formation is often the result, as seen in the large ring of star formation that surrounds the galaxy to the left of the two smaller galaxies. This Hubble Space Telescope *image shows the famous "cartwheel" galaxy, and the escaping culprit from the recent hit-and-run. One of the two smaller galaxies to the right in the image on the right is the guilty party, and the ring surrounding the cartwheel galaxy is teeming with young massive stars.*

(Image from NASA)

A Piece of the Action

From what we have seen so far of galaxies, there aren't many slackers. They all seem to get a lot done on an average day. That is, they all seem quite active.

But "active galaxy" has a specific meaning to an astronomer. Indeed, what astronomers call active galaxies might well be an intermediate evolutionary stage between quasars and normal (or should we say mature?) galaxies.

Between the great distances to quasars and the more moderate distances to our local galactic neighbors are a vast number of normal galaxies, and a few galaxies that are more luminous (particularly in their central or nuclear region) than average. It is these latter objects that astronomers call *active galaxies.*

The excess luminosity tends to be concentrated in the nucleus of the galaxy, and the centers of these galaxies are referred to as active galactic nuclei. As a class, active galaxies have bright emission lines (implying that their centers are hot), or are variable on short time scales. Some of them also have jets of radio emissions emanating from their centers, stretching hundreds of thousands of light-years into intergalactic space.

Star Words

Active galaxies are galaxies that have more luminous centers than normal galaxies.

The Violent Galaxies of Seyfert

In 1944, Carl K. Seyfert, an American astronomer, first described a subset of spiral galaxies characterized by a bright central region containing strong, broad emission lines. These *Seyfert galaxies* have luminosities that vary, and some show evidence of violent activity in their cores.

Here are a few distinguishing characteristics of Seyfert galaxies:

♦ Spectra emitted by Seyfert nuclei have broad emission lines, which indicate the presence of very hot gas or gases that are rotating at extreme velocities.

♦ The radiation emitted from Seyfert galaxies is most intense at infrared and radio wavelengths and must be, therefore, nonstellar in origin.

♦ The energy emitted by Seyfert galaxies fluctuates significantly over rather short periods of time.

> **Astro Byte**
>
> Only one percent of all spiral galaxies are classified as Seyfert galaxies. Here's another way to think about this: Perhaps all spiral galaxies exhibit Seyfert properties 1 percent of the time.

At the heart of all Seyfert galaxies is something relatively small but extremely massive, massive enough to (periodically) create the tremendous activity evident at the Seyfert galaxy nucleus. One likely possibility? The core of a Seyfert galaxy might contain a massive black hole.

Cores, Jets, and Lobes: Radio Galaxy Anatomy

Another kind of active galaxy is called a *radio galaxy*. Although Seyferts are an active subclass of spiral galaxies, radio galaxies are an active subclass of elliptical galaxies.

> **Star Words**
>
> **Radio galaxies** are an active galaxy subclass of elliptical galaxies. They are characterized by strong radio emissions, and in some cases, narrow jets and wispy lobes of emissions located hundreds of thousands of light-years away from the nucleus.

There are many types of radio galaxies, often classified by their shapes. Some radio sources have emissions only in their nucleus. In others, two narrow streams—or jets—of oppositely directed radio emissions emerge from the galaxy nucleus. The jets in these so-called double radio sources often end in wispy, complex puffs of radio emissions much larger than the central elliptical galaxy. These are called radio lobes. In some radio galaxies lobe emissions dominate, and in others jet emissions dominate.

One remarkable aspect of the jets of radio emissions is that they are observed over such a huge range of scales. The jets in some galaxies are linear for hundreds of thousands of light-years, and yet are observable down to the smallest scales that we can see—a few light-years. Radio jets are thought to be beams of ionized material that have been ejected from near the galactic center. The jets eventually become unstable and disperse into lobes.

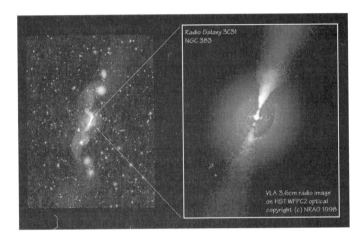

The radio galaxy 3C31 is shown superimposed on an optical image of the same region. In the inset to the right, the VLA image (jets) is superimposed on a Hubble Space Telescope *image of the region. When world-class telescopes like the VLA and the* HST *improve their resolutions in sync, useful overlays like these are possible.*

(Image from NRAO/NASA)

Close Encounter

Many differences in astronomy come down to a question of perspective. Astronomers are not entirely certain that core-halo radio galaxies (with a bright center and a diffuse envelope of emissions) and lobe radio galaxies (with a distinct bright center and diffuse lobes) are unique objects. That is, a core-halo galaxy might be nothing more than a foreshortened view of a lobe radio galaxy. If the galaxy happens to be oriented so that we view it through the end of one of its lobes, it will look like a core-halo galaxy. If we happen to see the galaxy from its side, two widely spaced radio lobes will be detected on either side of a central core.

Material in radio jets is being accelerated to enormous velocity. In some radio jets, bright blobs of radio emissions appear to be moving faster than light. And that's not possible!

What's going on? The apparent *superluminal motion* (speed faster than light) results from jets that are moving toward us, giving the illusion of impossibly high velocity. It does not defy any laws of physics.

Star Words _____

Superluminal motion is a term for the apparent "faster than light" motion of the blobs in some radio jets. This effect results from radio-emitting blobs that are moving at high velocity toward us, so that the radio emissions seem to be impossibly accelerated.

Star Words _____

Synchrotron radiation arises when charged particles are accelerated by strong magnetic fields. The emissions from radio galaxies are mostly synchrotron.

What is this mysterious *radio emission* we have been discussing? Well, radio emissions come in two basic flavors. One is rather bland, and the other a bit more spicy. The bland radio emission is *thermal emission*, which arises most commonly in regions of hot, ionized gas like the HII ("H-two") regions around young, massive stars. This type of radio emission comes from free electrons that are zipping around in the hot gas.

The spicy variety is *synchrotron emission*, sometimes called *nonthermal emission*. This type of radiation arises from charged particles that are being accelerated by strong magnetic fields. (James Clark Maxwell first discussed the effects of magnetic fields on charged particles.) The intensity of synchrotron radiation is not tied to the temperature of the source, but to the strength of the magnetic fields that are there. Radio jets, filled with charged particles and laced with strong magnetic fields, are intense sources of *synchrotron radiation*.

Where It All Starts

Quasars and active galaxies are sources of tremendous energy. We have seen that stars cannot account for their energy output, and that most of their energy arises from a small region at the center of the galaxy. How can we account for the known properties of active galaxies and quasars, including their staggering luminosities?

We saw in the last chapter that the rotation curves of many galaxies show evidence of large mass accumulations in the central regions. In fact, the masses contained are so large and in such a small volume that astronomers have in many cases concluded that a black hole must be present at the galaxy's center.

If a black hole were present in the center of active galaxies (along with the quasars that they might have evolved from), many of the observed properties of these strange galaxy types could be explained. To begin with, the tremendous *nonstellar* energy output originating in a compact area points to the gravitational field and accretion disk of a black hole.

But not just any black hole.

These black holes dwarf the stellar mass black holes that we discussed in Chapter 15, and are called supermassive black holes.

The center of our own Galaxy, the Milky Way, might contain a black hole with a million (or more) solar masses. To account for the much greater luminosity of an active galaxy, the mass must be much higher—perhaps a billion solar masses.

If black holes are the engines, we can explain the luminosity of active galaxies and quasars as the result of gas that spirals toward the black hole at great velocity, becoming heated in the process and producing energy in the form of electromagnetic radiation. X-ray photons are often produced by this process, and these are visible with the new *Chandra X-ray Observatory.*

In the black hole model, the radio jet arises when the hot gas streams away from the accretion disk in the direction of least resistance—perpendicular to the accretion disk. These jets, then, can stream away from the disk in two directions, giving rise to the oppositely directed jets.

> ### Astronomer's Notebook
>
> Remember the fluctuations in brightness seen in Seyfert galaxies and quasars? These fluctuations might be explained by brightness fluctuations in the accretion disk—the swirling disk of gas spiraling toward the black hole.

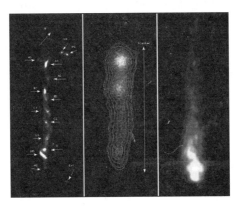

This image shows the jet associated with the quasar 3C 273. On the left is the optical image (from the Hubble Space Telescope*). In the middle is the Chandra X-ray Observatory image, and on the right is the radio image of the jet as seen by the* Multi-Element Radio Linked Interferometer Network *(MERLIN). Notice that the different wavelengths emphasize different parts of the jet.*

(Image from NASA, Chandra and Merlin)

If black holes exist at the centers of all or most galaxies, then the differences between quasars, radio galaxies, Seyfert galaxies, and normal galaxies is not the engine, but the fuel. When fuel is plentiful (as it is early in the universe), the cores of all galaxies might burn bright as a quasar. Later, they would move through a quieter, yet still active phase as Seyfert or radio galaxies. Later still, at our epoch in the universe, with fuel less plentiful, most black holes lie dormant. Models of galaxy evolution continue to be hotly debated (at least by astronomers), but the existence of central black holes might answer many questions.

A closing thought, then. Let's say you take the trouble to find the quasar 3C 273 for yourself. You see a quasar—but the light that you see left the galaxy some 2 billion years ago. The quasar you see might now (*now* as it is experienced at the distant location of the 3C 273) be a more mature galaxy, not unlike the Milky Way. Perhaps on some planet orbiting an average star somewhere in the distant universe, an amateur astronomer is pointing a telescope at our Milky Way, which appears to him or her or it as a faint, bluish blob of light: a quasar.

The Least You Need to Know

- Quasars were first discovered at radio wavelengths and are some of the most distant astronomical objects visible.

- Quasars might be the ancestors of all galaxies, the violent beginnings of us all.

- Active galaxies are any galaxies that are more luminous than what we call normal galaxies, with bright starlike cores and broad, strong emission lines.

- Seyfert galaxies are the active subset of spiral galaxies, and radio galaxies are the active subset of elliptical galaxies.

- Radio jets originate at the cores of active elliptical galaxies and terminate in wispy patches of emissions called lobes.

- Supermassive black holes are the most likely source of energy for quasars and active galaxies.

Part 5

The Big Questions

The questions themselves might be expressed simply enough: Is the universe eternal? If not, how did it begin? And how will it end? Is the universe infinite? Finally, in this universe, are we alone? Is there life—other than on Earth—within the solar system? The Galaxy? Beyond? If so, is it intelligent? And if intelligent, will it communicate with us?

The questions are put simply enough. The answers … well, read on.

Chapter 19

Cosmology and Cosmologies

In This Chapter

- ◆ What a licensed cosmologist does (Hint: it doesn't involve makeup)
- ◆ Understanding the cosmological principle
- ◆ All about the Big Bang
- ◆ Critical density and the expanding universe
- ◆ Matter from energy and energy from matter

Cosmology. The word has an archaic ring to it. No wonder, because it's been part of the English language since at least the 1600s, and because it describes the study of some of the oldest and most profound issues humankind has ever addressed: the nature, structure, origin, and fate of the universe. Those are some big questions.

For centuries, cosmology was the province of priests and philosophers. This chapter shows what happened when astronomers took up the subject in the twentieth century.

The Work of the Cosmologist

Many of us grew up thinking that the universe was both infinite in size and eternal in time. The planets might move in the sky, and, farther away, stars might orbit the centers of galaxies. But on the largest scales, certainly the universe must be changeless. It has always been and it will always be. Right? How, after all, could it be otherwise?

Actually, long before *cosmologists* came on the scene to study the origins of the universe, there were creation stories, mythological and religious narratives that explained the origin of the universe. In the Judeo-Christian tradition, for example, God said, "Let there be light." And there it was. In Hinduism there are diverse beliefs about the origin of the universe, but many believe the universe cyclically recreates itself once all *karma* is extinguished. Some Buddhists believe that the world recreates itself every tiny fraction of a second.

> **Star Words**
>
> **Cosmology** is the study of the origin, structure, and fate of the universe.

The human mind naturally looks for origins, beginnings, and grand openings. In the twentieth century, humans stumbled across two important bits of evidence of the biggest grand opening ever.

> **Astronomer's Notebook**
>
> For decades, astronomers had to struggle with cosmological questions based on a bare minimum of observational evidence. The past two decades have provided observers and theorists with a wealth of new information about what the universe looked like in its infancy. These observational clues have triggered an explosion in our understanding of the origin of the universe.

Modern cosmologists have put together a creation story of their own. It is the result of centuries of observation, modeling, and testing. The model is so well established that it is generally referred to as the *standard model*. As Steven Weinberg points out in his account of the start of it all, *The First Three Minutes* (Basic Books, 1993), the study of the origins of the universe was not always a scientifically respectable pursuit. It is only in relatively recent scientific history that we have had observational tools rigorous enough to test theories of how the universe began.

Two New Clues

The twentieth century threw some major curve balls at the human psyche, from the carnage of two world wars to the invention of the hydrogen bomb, capable of releasing the fury of a stellar interior on the surface of our planet. And although the seventeenth century saw Earth pushed from the center of the universe, it was early twentieth-century astronomers who told us that we were located in the suburbs of an average galaxy, hurtling away from all the other galaxies that we can see.

Why are we all flying away from one another? Because the universe exploded into being about 14.5 billion years ago!

Redshifting Away

Recall from Chapter 17 our discussion of Hubble's Law and Hubble's constant. We know from observing the redshift in the spectral lines of galaxies that all the galaxies in the universe are flying away from us, and that the galaxies farthest away are moving the fastest. And any observer in any galaxy would see exactly what we do.

Armed with Hubble's Law (and a value for the Hubble constant), it is easy to calculate how long it has taken any given galaxy to reach its present distance from us. In a universe that is expanding at a constant rate, we simply divide the distance by the velocity. The answer for all the galaxies we see is about 14.5 billion years.

So if we were able to rewind the expanding universe to see how long it would take for all the galaxies that are flying apart to come together, the answer would be about 15 billion years.

Pigeon Droppings and the Big Bang

The Russian-born American nuclear physicist and cosmologist George Gamow (1904–1968) was the first to propose (in the 1940s) that the recession of galaxies (known about since the 1920s) implied that the universe began in a spectacular explosion. It was not an explosion *in* space, but an explosion *of all* space. The entire universe was filled with the explosion that started the universe.

Later, British mathematician and astronomer Sir Fred Hoyle (born 1915) coined the term *Big Bang* for this event. He did not mean it as a compliment. With astronomers Herman Bondi and Thomas Gold, he proposed an alternative *steady state theory* of the universe, proposing that the universe is eternal, because it had no beginning and will have no end.

Quite soon after Gamow proposed the Big Bang theory, theorists realized that the early universe must have been incredibly hot, and that the electromagnetic echoes of this colossal explosion should be detectable as long-wavelength radiation that would fill all space. This radiation was called the *cosmic microwave background*.

You might recall from Chapter 5 that the science of radio astronomy was the accidental by-product of a telephone engineer's search for sources of annoying radio interference. Another radio-frequency accident resulted in an even more profound discovery. From 1964 to 1965, two Bell Telephone Laboratories scientists, Arno Penzias and Robert Wilson, were working on a project to relay telephone

Star Words

The **Big Bang** refers to the origin (in space and time) of the universe. Cosmologists propose that the universe started its life in an incredibly dense and hot state.

calls via early satellites. Wherever they directed their antenna on the sky, however, they detected the same faint background noise. The noise was not associated with any particular point on the sky, but was the same everywhere. In an effort to remove all possible sources of interference, they went so far as to scrape off the pigeon droppings they found on the antenna because these were a potential source of heat (energy, which could be a source of radio interference). But even when their antenna was clean as a whistle the noise persisted.

Star Words

The **cosmic microwave background** consists of the highly redshifted photons left behind by the Big Bang and detectable today throughout all space as radio-wavelength radiation.

The "noise" they had stumbled across was the very *cosmic microwave background* that had been predicted, the highly redshifted photons left behind by the Big Bang. The theory goes like this: Immediately after the Big Bang, the energy liberated was most intense in the shortest, highest-energy wavelengths—gamma rays. As the universe expanded and cooled over millions, then billions of years, the waves expanded as well, reaching us today as long-wavelength photons in the microwave part of the electromagnetic spectrum.

This microwave radiation carries information from the early moments of our universe. It tells us that there is a remnant echo of the Big Bang found permeating the universe, evidence that in past times the universe was a much hotter place than it is now. The *Cosmic Background Explorer* (COBE) satellite was launched in November 1989. In the first few minutes of its operation, it measured the temperature of the background radiation to be 2.735 K.

The COBE Differential Microwave Radiometer (DMR) found variations in the cosmic microwave background (shown here as brightness variations) above and below a temperature of 2.73 K. The variations are tiny—at the level of a few parts in a million.

(Image Courtesy of NASA/COBE)

This satellite improved greatly on the early, crude mappings of the cosmic microwave background. In addition, we have also learned from *COBE* that the solar system has a small motion relative to the cosmic microwave background. Finally, in 1992, *COBE* data revealed yet another exciting fact: There are tiny fluctuations on the sky in this background radiation; it is not exactly homogeneous, or smooth. These temperature differences are the result of density fluctuations in the early universe, which appear to have seeded the universe for the eventual formation of galaxies.

Same Old Same Old

Edwin Hubble also made two very important discoveries in his studies of galaxy types and distributions. He found that the universe appeared to be both isotropic (the same in all directions), and homogeneous (one volume of space is much like any other volume of space).

His conclusion was the result of painstaking study of the distribution and type of a large number of galaxies at various redshifts, or distances, from Earth.

The Cosmological Principle

Together, the homogeneity and isotropy of the universe make up the *cosmological principle:* one of the cornerstone assumptions of modern cosmology. If we could not make this assumption (based on observation), then our cosmology might only apply to a very local part of the universe.

But the cosmological principle enables us to extrapolate conclusions drawn from our local viewpoint to the whole universe. Consider these important implications: A homogeneous universe can have no border or edge (because it is the same in any volume), nor can it have a center (because it should look the same in all directions from any viewpoint).

Perhaps the most familiar model of the expanding universe is a toy balloon, on which we randomly draw some dots with a magic marker. Magic-marker dots on the balloon's surface represent galaxies. But please heed this warning first: If the universe is represented by the surface of a balloon, then we are talking about a two-dimensional universe. There is no inside of the balloon if you can draw only on its surface. Nevertheless, we need a two-dimensional model of our three-dimensional universe to picture the expansion.

Star Words

The **cosmological principle** is a cornerstone assumption (based on observation) about the nature of the universe. It holds that the universe exhibits two key properties: homogeneity (sameness of structure on the largest scales) and isotropy (it looks the same in all directions).

If you, as a two-dimensional being, look out into your two-dimensional universe (out into the balloon's surface), you see no center and no edge, and if the balloon is inflated, the dots move away from one another. Regardless of their point of view, all the dots "see" all the others as moving away.

This balloon model, by the way, represents a "closed" universe, which we will talk about in Chapter 20. In an infinite and "flat" universe, we would represent the universe with an infinitely large rubber sheet, which we stretched in all directions.

Big Bang in a Nutshell

You might be sitting there thinking, "But what caused the Big Bang?"

That is a question mostly outside the bounds of astronomy and physics. Scientists can hope only to understand things that are of this universe, and whatever caused the Big Bang is inherently unobservable. Nevertheless, we will return to this question briefly in Chapter 21.

Universal recession in time means the universe is expanding, and expansion, in turn, implies an origin time. Astronomers believe that at its origin, the universe was unimaginably dense. The entire universe was a point, with nothing outside the point. Almost 15 billion years ago, the entire universe exploded in the Big Bang, and the universe has been expanding ever since, cooling and coalescing into ever more organized states of matter. One very highly organized state of matter (you) is reading this book. Hubble's Law describes the rate of the observed expansion.

Big Bang Chronology

The prehistory of the Big Bang is inherently unknowable, but we are able to discuss what happened very soon *after* the Big Bang. And we *do* mean *very* soon. From about $\frac{1}{100}$ of a second after the Big Bang onward, we can outline the major steps in the evolution of the universe. It is basically a story of cooling and expanding. For a thorough review of modern cosmology, see Timothy Ferris's excellent and humorous *The Whole Shebang* (Touchstone, 1998). An expanded and more technical take on the early moments of the universe can be found in Steven Weinberg's *The First Three Minutes* (Basic Books, 1993).

At the earliest times that we can track, the universe was incredibly hot (10^{11} K) and filled with elementary particles, the building blocks of atoms: electrons, positrons, neutrinos, and photons of light. And for every billion or so electrons present at this time, there was one heavy particle (a neutron or proton).

In this soup—an entire universe hotter than the core of a star—energy and matter were going back and forth as electrons and positrons *annihilated* to produce energy, and more were born from the energetic photons that filled the universe. At about one second into the life of the universe, the temperature everywhere was 10^{10} K, still too hot for neutrons and protons to be bound into nuclei.

As the universe rapidly cooled (after a mere 10–15 seconds), there was insufficient energy to create new electrons and positrons, so most of them annihilated, without new ones taking their place. It was cool enough at this point for helium nuclei (two protons and two neutrons) to form, but they could only form after deuterium (an isotope of hydrogen consisting of one proton and one neutron) formed. The instability of deuterium at this temperature means that the universe still couldn't form much helium. After about the first three minutes of its existence, the universe had cooled to some one billion degrees. Most of the electrons and positrons had annihilated, and the universe then consisted of photons of light, neutrinos and antineutrinos, and a relatively small amount of nuclear material. At about this time, deuterium nuclei were able to hold together, and the entire universe (for a moment) acted like the core of a star, fusing helium and small amounts of lithium in a phase called *primordial nucleosynthesis.*

Star Words

Particles and antiparticles **annihilate** when they meet, converting their mass into energy. Electrons and anti-electrons (positrons) were continually created and annihilated in the early universe.

The universe was still far too hot for electrons to come together with nuclei to form stable atoms. Not until about 300,000 years later (when the entire universe was at the temperature of a stellar photosphere) would nuclei be able to hold on to electrons.

It wasn't until after this first 300,000 years that the universe became transparent to its own radiation. What does that mean? Before the electrons settled out into atoms, they got in the way of all the photons in the universe and kept colliding with them. With the electrons cleared out of the picture (grabbed by nuclei), the universe shifted from a *radiation-dominated* state to a *matter-dominated* state.

Here's another way to think about this idea: For the first 300,000 years after the Big Bang, the universe was "foggy" with electrons. This "fog" of electrons made the universe opaque to its own photons. When the universe was cool enough for electrons and nuclei to combine, it suddenly became transparent. The cosmic microwave background comes from these first freed photons.

From this point on, the universe continued simply to cool and coalesce, eventually forming the stars and galaxies of the observable universe.

A Long Way from Nowhere

Now just *where* did this Big Bang occur, you might be wondering?

If the Big Bang took place *somewhere*, that place is, by definition, the center of the universe. Yet the cosmological principle forbids any such center. The Big Bang might resolve some open questions, but, it would seem, it also torpedoes the cosmological principle.

Well, actually, it doesn't—as long as we conclude that the Big Bang took place *everywhere*. And the only way such a conclusion could be true is if the Big Bang was not an explosion within an empty universe—the way a movie explosion is—but was an explosion of the universe itself. At early times in the universe, its entire volume was hotter than a stellar interior. The universe was and is all that exists and has ever existed.

Looked at this way, it is a symptom of geocentric bias to say that the galaxies are receding from us. Rather, the universe itself is expanding, and the galaxies (ourselves included) are moving along with that expansion.

Astro Byte

At the moment of the Big Bang, the entire universe came into being. As the universe cooled, the elements hydrogen (73 percent) and helium (27 percent) were created in the primordial fireball, along with small amounts of lithium.

How Was the Universe Made?

Radiation dominated over matter in the early universe. Although photons (radiation) still outnumber atoms (matter) by about a billion to one in the universe, matter now contains far more energy than radiation. A simple calculation, using Einstein's celebrated equation showing the equivalence of energy and mass, $E=mc^2$, demonstrates that matter, not radiation, is dominant in the current universe. The energy contained in all the mass in the universe is greater than the energy contained in all the radiation.

There is a twist here, however, which we will explore in Chapter 21. The entire universe might be dominated by "dark energy" driving its continued expansion.

As the universe expands, both radiation and matter become less concentrated; they travel with the expansion. The energy of the radiation is diminished more rapidly than the density of matter, however, because the photons are redshifted, becoming less energetic as their wavelength lengthens. Thus, over time, matter has come to dominate the universe.

Mommy, Where Do Atoms Come From?

When we looked at fusion in the cores of stars (see Chapters 12 and 15), we noted that hydrogen served as the nuclear fuel in a process that produced helium. Yet there is so much helium in the universe—helium accounts for more than 23 percent of the mass of the universe—that stellar fusion cannot have produced it all. It turns out that most of the helium in the universe was created in the moments following the Big Bang, when the universe had cooled sufficiently for nuclei to hold together, after about the first three minutes.

Hydrogen fusion in the early universe proceeded much as it does in the cores of stars. A proton and a neutron come together to form deuterium (an isotope of hydrogen), and two deuterium atoms combine to form a helium nucleus.

Careful calculations show that between 100 and 120 seconds after the Big Bang, deuterium would be torn apart by gamma rays as soon as it was formed. However, after about two or three minutes, the universe had sufficiently cooled to allow the deuterium to remain intact long enough to be converted into helium. As the universe continued to cool, temperatures fell below the critical temperature required for fusion, and *primordial nucleosynthesis* ended.

Only after about 300,000 years had passed would the universe expand and cool sufficiently to allow electrons and nuclei to combine into atoms of hydrogen and helium. It is at this point, with the temperature of the entire universe at some 5,000 K, that photons could first move freely through the universe.

The cosmic microwave background we can now measure consists of photons that became free to move into the universe at this early time.

The rest of the periodic table—the other elements of the universe—beyond lithium would be filled out by fusion reactions in the cores of stars and supernova explosions of massive stars.

Star Words

Primordial nucleosynthesis refers to the fusion reactions that occurred in the early universe at temperatures of about 10^{10} K. These fusion reactions produced mostly helium and small amounts of lithium.

Astro Byte

When the entire universe was the temperature of a cool stellar photosphere (about 5,000 K), the electrons were able to join together with nuclei, forming hydrogen and helium atoms and causing the universe to become

The cosmic microwave background (CMB) can be very well modeled with a black body that has a temperature of 2.725 K (curve shown here).

(Image Courtesy of NASA)

SPECTRUM OF THE COSMIC MICROWAVE BACKGROUND

MAP990045

Stretching the Waves

As mentioned earlier, there is another way to think about the observed redshift in galaxy spectra. The photons emitted by a receding galaxy are like elastic bands in the fabric of the universe: They expand with it. In this sense, the redshift we measure from distant galaxies is nothing less than a *direct* measurement of the expansion of the universe.

We have seen where the universe and all that is within it have come from. Now we turn finally to where it is all going and where it will end. Will the universe expand forever? Or will it turn back in on itself and end in a final collapse? As it turns out, it all depends on gravity.

Depending on how much mass there is in the universe, it will either expand forever or collapse back in on itself in the opposite of a Big Bang: a Big Crunch.

In Chapter 21, we will discuss recent observations of high-redshift supernovae and the cosmic microwave background that suggest a Big Crunch will never happen. The universe appears destined—or doomed—to expand forever.

The Least You Need to Know

◆ The redshift of galaxies and the presence of a cosmic microwave background are two clues that the universe had a beginning in time.

◆ The cosmological principle holds that the universe is homogeneous (uniform in structure on large scales) and isotropic (it looks the same in all directions). This principle allows us to generalize from local observations.

◆ The current theory of the origin of the universe is the Big Bang, which holds that the universe began as an explosion that filled all space.

◆ For the first 300,000 years after the Big Bang, the universe was "foggy" with electrons, and opaque to its own photons. When the universe was cool enough for electrons and nuclei to combine, it became transparent. The cosmic microwave background comes from these first freed photons.

◆ All matter in the universe was created at the moment of the Big Bang. Since that time, stellar fusion and supernovae have converted the dominant hydrogen and helium into all of the elements of the periodic table.

Chapter 20

(How) Will It All End?

In This Chapter

- ◆ The future of the universe: expansion or contraction?
- ◆ How the density of the universe determines its fate
- ◆ The universe before the Big Bang
- ◆ The shape of the universe
- ◆ Perfecting the Big Bang theory

In Woody Allen's 1977 movie *Annie Hall*, there is a flashback to the Brooklyn childhood of the main character, a stand-up comic named Alvy Singer. The family doctor has been summoned because young Alvy, perhaps 10 years old, is depressed. He has just learned that the universe is expanding. He believes it will eventually fly apart. The physician (puffing on a cigarette) offers the comforting thought that universal calamity is many billions of years in the future. Alvy's mother is more strident: "What is it your business? We live in *Brooklyn*. Brooklyn is *not* expanding!"

In this chapter, we strive to be more precise than both Alvy's doctor and his mother. We look toward the end of the universe—if there will be an end. Will it end with a bang or a whimper? Will it end with fire or with ice?

What the Redshift Means

Alvy's mother was right. The truth is that Brooklyn itself *isn't* expanding. But the universe that revolves around it (we're taking a Brooklynite's perspective here) is expanding. Much to the chagrin of New Yorkers, Brooklyn is not somehow exempt from the laws of the universe. This borough—like Earth, like a star, like a galaxy, or like a human being—is held together by its own internal forces. The expansion of the universe that is the echo of the Big Bang is only visible on the largest scale, the distances between clusters and superclusters of galaxies.

Let's return to our picture of the universe as the surface of a rubber balloon. As the balloon is inflated, the distance between any two dots drawn on its surface increases. The resulting observation is that these receding galaxies emit photons that are redshifted. Another way to think of it is that the photons that leave another galaxy are part of the fabric of the universe, and as this fabric is stretched, each photon gets stretched along with it—to a longer (redder) wavelength. This type of wavelength increase is called a *cosmological redshift*, and it is a direct measure of universal expansion.

Star Words

Cosmological redshift is the lengthening of the wavelengths of electromagnetic radiation caused by the expansion of the universe.

Limited Options

Please keep this in mind: The values we measure locally, related to the expansion of the universe, are not necessarily "universal" in time That is, just because the universe is expanding at a certain rate now in our particular neighborhood doesn't mean that it has always expanded at that rate. It might have been expanding more slowly, or more quickly in the distant past. The only way to find out is to measure the *redshift* of very distant objects.

And just because the universe is expanding now doesn't necessarily mean that it will go on expanding forever. The expansion that presumably started with the Big Bang is opposed by a relentless force you have met many times before in this book: *gravity*. The eventual fate of the universe hangs on how much stuff (mass and energy) there is in the universe. The density of matter and energy in the universe will determine whether it ends in fire or in ice.

There are basically three options: The universe will either expand forever (be unbound), or it will continually slow in its expansion but never stop (be marginally bound), or it will reach a certain size and then begin to contract (be bound).

Close Encounter

Some recent discoveries have a bearing on the question of the eventual fate of the universe. Two groups of astronomers (one at the Harvard-Smithsonian Center for Astrophysics, and the other at Berkeley Labs) using the *Hubble Space Telescope* have found evidence that the universe is likely to expand forever, that the universe doesn't have enough mass in it to cause the expansion to halt.

They have used what are called Type Ia supernovae (cataclysmic explosions that result when a white dwarf star borrows a bit too much material from its binary companion) as standard candles. Type Ia supernovae are so incredibly luminous that they enable us to see farther than just about any other phenomenon we know of. Using these supernovae, astronomers have determined that the expansion of the universe (as described by Hubble's Law) doesn't seem to be slowing down at all. In fact, it appears to be speeding up.

A Matter of Density

The origin of these limited options is the density of the universe. If the universe contains a sufficient amount of matter and energy—is sufficiently dense—then the force of gravity will be such that it will eventually succeed in halting expansion. The universe will gradually expand at a slower and slower rate, and then reverse itself and begin to collapse. If the universe (even counting all of the material that we can't see—the "dark matter") does not contain sufficient mass—is too thin—then the expansion of the universe will continue forever.

Remember that in the Big Bang theory, there is no such thing as new or old matter. All the matter there is was created as a result of the Big Bang; therefore, the density of the universe—and its eventual fate—was determined at the moment the Big Bang occurred.

There is a certain density of matter and energy in the universe that is the sword edge between ultimate collapse and eternal expansion. This *critical density* has been calculated. If the universe is more dense than this, it will eventually collapse; if it is less dense than this, it will expand forever; and if it is exactly this dense, it will keep expanding, more and more slowly, forever.

Star Words

Critical density is the density of matter and energy in the universe that represents the division between a universe that expands infinitely (unbound) and one that will ultimately collapse (bound). The density of the universe determines whether it will expand and get colder forever, or end with a conflagration as dramatic as the Big Bang.

Stalking the Wily Neutrino

Neutrinos were defined in Chapter 14 as subatomic particles without electric charge and with virtually no mass. As a result, these are sort of stealth particles: small, neutral particles that are hard to detect. On June 5, 1998, U.S. and Japanese researchers in Takayama, Japan, announced that, using an enormous detection device, they found evidence of mass in the neutrino. The device, called the Super-Kamiokande, is a tank filled with 12.5 million gallons of pure water and buried deep in an old zinc mine.

Because neutrinos lack an electric charge and rarely interact with other atoms, passing unnoticed through virtually any kind or thickness of matter, they have proven especially elusive. They are the greased pigs of the subatomic world, so plentiful that 100 billion of them (that's as many stars as there are in the Galaxy) pass through your body every second. The Super-Kamiokande contains so much water, however, that occasionally a neutrino collides with another particle and produces an instantaneous flash of light, which is recorded by a vast array of light-amplifying detectors.

So if the researchers are right and neutrinos do have mass, just how much mass do they have? It doesn't appear to be enough to affect critical density.

The Super-Kamiokande detector began observations in 1996, and consists of a total of 50,000 tons of water. There are outer and inner volumes, the inner volume ringed with 11,200 photomultiplier tubes (PMTs), seen here during maintenance. The tubes are sensitive to light called "Cerenkov light" which is emitted by high velocity particles traveling in the water.

(Image courtesy of U. Tokyo/UC Irvine/BU)

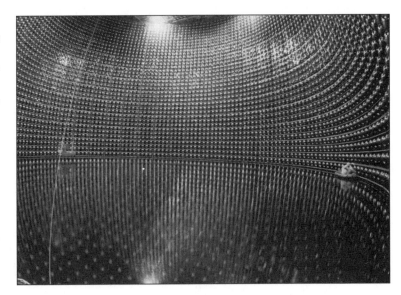

Run Away! Run Away!

To predict whether the universe will expand infinitely, it is necessary to calculate critical density and to see whether the density of our universe is above or below this critical figure. By measuring the average mass of galaxies within a known volume of space, astronomers derive a density of luminous matter well short of critical density, about 10^{-28} kg/m³. That figure is about 1 percent of what is required to "close" the universe. After factoring in the added contribution of the dark matter that has an apparent gravitational effect in galaxy clusters (see Chapter 16), we can account for up to about 30 percent of the critical density.

Is the universe, then, far from critical density? On the contrary, recent experiments suggest that it is very close to critical density and that the "missing" 70 percent is not made up of mass, but of energy. Remember that energy and mass can be interchanged through the process described by $E = mc^2$. If the expansion of the universe is accelerating, and the universe has "critical density," these might both be explained by energy contained in the vacuum of space.

What Does It All Mean?

You should now start to see the outlines of the different possible fates for the universe. At present, the evidence is starting to point toward an open, or unbound, universe—one that will expand forever, growing ever colder and more diffuse with time. Eventually, all the stars in the universe, even the low-mass ones, will have turned off, and the universe will be a dark, cold place indeed. But before we throw in the towel, let's consider what some of the possible fates of the universe might mean for us.

What's the Point?

Remember your high school geometry class? You were told that a point is a fiction, represented by the mark of a very sharp pencil, perhaps, but actually without dimension: no length, no height, no width. Are cosmologists asking us to accept that this *point* once contained the entire universe and that it might contain it again? Not quite. The point cosmologists speak of is not the universe *within* something else, some void. Their point is not one within space. Rather, it contains the entire universe. Everything that ever is or was in the universe was there.

The Universe: Closed, Open, or Flat?

So then the universe expanded from this point in the Big Bang. What was the eventual result? What is the geometry of the universe?

To begin to visualize the shape of the universe, we have to stop thinking like Newton and start thinking like Einstein. Remember that Einstein thought of gravitation not as a force that objects exert upon one another, but as the result of the distortion in space that mass causes. Einstein explained that the presence of mass warps the space in its vicinity. The more mass (the greater the density of matter), the greater the warp (the more space curves).

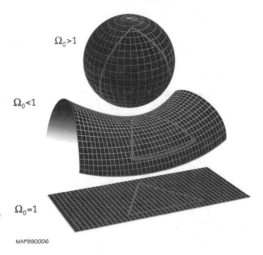

$\Omega_0>1$

$\Omega_0<1$

$\Omega_0=1$

MAP990006

The universe might be "closed," "open," or "flat," depending on whether the universe is at critical density or not. This image shows two-dimensional models that we use to understand these three basic geometries. A closed universe (top figure) can be thought of as curved like the surface of a sphere, and finite in extent. An "open" universe (middle figure) can be thought of as saddle-shaped and infinite in extent. A "flat" universe (bottom figure) can be thought of as a plane, also infinite in extent.

(Image courtesy of NASA)

If the density of the universe is greater than the level of critical density ($[om]_o>1$), then the universe will warp (or curve) back on itself, closed and finite. If the universe is closed, then our balloon analogy has been particularly accurate. In a *closed universe*, we think of the entire universe as represented by the surface of a sphere, finite in extent, but with no boundary or edge. If you shot out parallel beams of light in a closed universe, they would eventually cross paths, like lines of longitude on a globe.

If the density of the universe is below the critical level ($[om]_o<1$), an *open universe* will result. Whereas the spherical or closed universe is said to be positively curved, the open universe is negatively curved.

Star Words

A **closed universe** is finite and without boundaries. A universe with density above the critical value ($[om]_o>1$) is necessarily closed. An **open universe** will expand forever, because its density is insufficient to halt the expansion ($[om]_o<1$).

Saddle Up the Horses: Into the Wide-Open Universe

The closest two-dimensional approximation of the open universe is the surface of a saddle, curved up in one direction and down in the other—and extending out to infinity. In an open universe, beams of light that were initially parallel would diverge. Our current measurements of the density of the universe suggest that this is the type of universe we inhabit.

With precisely critical density, by the way, the universe becomes *flat*, albeit still infinite in extent. In what we call flat space, parallel beams of light would remain parallel forever. For this reason, flat space is sometimes called Euclidean space. Recent observational results (explored in Chapter 21) suggest that the universe might be flat.

Star Words

A **flat universe** results if density is precisely at the critical level ($[om]_o=1$). It is flat in the sense that its space is defined by the rules of ordinary Euclidean geometry—parallel lines never cross.

We Have a Problem

Scientists are to their theories as overbearing parents are to their children. They have for them only the very highest expectations. Contradictions are not allowed and must be resolved. Even something less than an outright contradiction is intolerable. A good theory should account for all observations and make testable predictions. Any observations in contradiction to the theory are taken seriously and the theory amended, if necessary.

In this regard, although it has been very successful, the Big Bang theory seems to fall short in two areas: explaining the incredible sameness of the universe on its largest scales and accounting for why the universe has a density, roughly 30 percent mass and 70 percent energy, that is apparently so close to critical.

First there is the so-called horizon problem, which deals with the incredible uniformity of the cosmic microwave background. No matter where we look, its intensity is the same, even in parts of the universe that are far too distant to be in contact with one another. Regions in the universe that never could have exchanged information (because of the limiting speed of light) seem to know about each other.

It is not that this uniformity contradicts anything in the Big Bang theory, it's just that there is nothing in the theory that accounts for it. The theory provides no particular reason why two widely separated regions should be the same—especially because regions very distant from one another would never have interacted.

The second issue is called the flatness problem. Its name comes from the fact that the critical density of the universe, once the mass of dark matter and the role of energy are figured into things, approaches 1, which means that the universe is almost flat.

So what's the problem?

Again, as with the horizon problem, the difficulty is not that flatness contradicts the Big Bang theory (it does not), but that the theory doesn't explain *why* the universe should have formed so close to critical density. As far as the theory goes, the universe might have been significantly more dense than 1 or significantly less. Scientists won't accept the luck of the draw. They want an explanation.

Close Encounter

There are four known forces in the universe: gravity, electromagnetism, the weak nuclear force, and the strong nuclear force. Gravity and the electromagnetic force act over large distances—the size of the universe—and govern the motions of planets and galaxies, molecules and atoms. The nuclear forces act over only small distances (within an atomic nucleus) and hold nuclei together. The nuclear forces are the strongest, but they act over the shortest distances. Gravity is by far the weakest, but its long reach means that its pull will govern the fate of the universe.

In the early universe, it is believed that these forces started as a single force. As the universe expanded, each force established its own unique identity.

Down to Earth

Successful people in all fields—from astronomy to retail sales—learn to see problems not as obstacles, but as opportunities. The biggest, most basic issues astronomy deals with seem the furthest removed from our everyday experience; but the possible resolution of the horizon and flatness problems might help us to connect more intuitively with this strange thing called the universe.

Blow It Up

Within the first infinitesimal fractions of the first second following the Big Bang, astronomers have theorized, the three forces other than gravity (electromagnetic, weak nuclear forces, and strong nuclear forces) in the universe were united as a single force.

Between 10^{-35} seconds and 10^{-32} seconds after the Big Bang, gravity had already split out as a separate force, but the other forces were still one. For an unimaginably brief instant, gravity pushed the universe apart instead of pulling it together. Theorists call this moment *inflation*, for obvious reasons. Within this *inflationary epoch* (not to be confused with the late 1970s), the universe expanded 100 trillion trillion trillion trillion (10^{50}) times.

Of course, 10^{-35} to 10^{-32} seconds seems to us instantaneous; but there was time—however brief—before this, between the Big Bang and the onset of the epoch of inflation. During this period, all parts of the universe were in communication with one another. They had ample time to establish uniform physical properties *before* the epoch of inflation pushed them to opposite sides of the universe.

As the early universe continued to expand, it did so at a rate faster than its constituent regions could communicate with each other. The universe, in effect, outran information, so that the most extreme regions have been out of communication with one another since 10^{-32} seconds after the Big Bang. Yet they share the properties they had at the very instant of creation. They continue to share these properties today. Thus, with the addition of inflation, the Big Bang can account for the horizon problem, and we've tied up one loose end of the theory.

Star Words

The **inflationary epoch** refers to a time soon after the Big Bang when the universe was puffed up suddenly, increasing its scale by a factor of 10^{50} in an instant. This inflation could account for the incredible sameness, or uniformity, of the universe, even in regions that (without inflation) could never have been in contact with one another.

Looks Flat to Me

That leaves us with the flatness problem. Why (in terms of the Big Bang theory) should the universe be at or near critical density?

If a mass of external evidence didn't exist, it would be almost impossible to convince anyone that Earth is round. After all, it *looks* flat.

After we accept that Earth is curved, however, we understand that it looks flat because the radius of the globe is very large and, therefore, the arc of the curve it describes is extremely gradual. Take a small portion of any arc, and it will look, for all practical purposes, flat. If Earth were very small—or we were very large, or very far away—the curvature of Earth's surface would be apparent on a routine basis.

The universe might have been curved at the instant of its creation. But in expanding 10^{50} times during the epoch of inflation, it became—*on the scale of the observable universe*—flat.

A flat universe is consistent with a universe whose density is exactly critical. If the universe were more dense, it would be positively curved; if less dense, it would be negatively curved. If the universe is truly flat, its density must be exactly critical.

As you will see in the next chapter, a variety of observational evidence is leading astronomers to believe that we live in a flat universe with critical density.

The Least You Need to Know

- The future of the universe is limited to two basic possibilities: Either it will expand forever or eventually collapse.

- The density of matter in the universe determines whether gravity will be able to halt the expansion of the universe that started with the Big Bang.

- Depending on the observed density of the universe, we say that it is open, closed, or flat. An open universe will expand forever, a closed universe will collapse on itself, and a flat universe will expand at a slower and slower rate.

- Inflation, a sudden expansion in the scale of the universe that occurred early on, might explain why the universe appears both flat and homogeneous.

21

Big Bang and Beyond

In This Chapter

- The *Cosmic Microwave Background Explorer (COBE)* mission
- Supernova light curves as standard candles
- The accelerating expansion of the universe
- The scale of fluctuations in the cosmic microwave background (CMB)
- What is dark energy?

You know that feeling when you've just about got everything figured out, and you are ready to kick back and relax for a while—and then you notice something that makes you realize that you still have a lot of work to do?

Perhaps you are almost finished with a huge reading assignment for a class, when you realize that you actually have another 100 pages to read. Or you have just finished mopping the kitchen floor, and your dog runs across it with muddy paws. For a few years there in the 1990s, it was starting to seem like many things about the origin of the universe, the Big Bang, and the Hubble Law were fitting together very nicely.

And then, as is often the case in astronomy, some surprising new observations were made that forced us to see the workings of the universe in a completely new way. Like those muddy dog tracks across the spotless linoleum, they let us know that we still had a lot of work to do.

Smooth or Crunchy?

In Chapter 19 we discussed the homogeneity and isotropy of the universe. All this means is that the universe appears to be the same over large volumes (homogeneity), and looks the same in all directions (isotropy). Launched in 1989, the *Cosmic Microwave Background Explorer (COBE)* satellite made observations of the cosmic microwave background (CMB) and confirmed that the universe was indeed isotropic. That is, to very high accuracy, after the motion of the solar system through the CMB was subtracted, the CMB was incredibly smooth.

Or was it?

In 1992, after more analysis of the data, the *COBE* team announced a surprising result. At very low levels, levels of one part in 100,000, there *were* fluctuations in the CMB. That is, the CMB appeared to have some structure. At very low levels, the universe was not isotropic!

COBE: Small Fluctuations

As the first instrument to measure the CMB from space, *COBE* certainly had some limitations, but there was no doubt that it could accurately measure the microwave temperature. Using the Differential Microwave Radiometer (DMR), the *COBE* team measured the "sky temperature" with amazing accuracy and found it to be 2.725 K (give or take 0.001 K).

This image from COBE *(the* Cosmic Microwave Background Explorer*) shows three views of the all-sky map of the cosmic microwave background (CMB). The top image shows combined emissions from the Milky Way, and our motion through the CMB. The middle image has had our motion (the dipole) removed and shows only the Galaxy's emissions. The map showing only small-scale fluctuations in the CMB is at the bottom.*

(Image courtesy of COBE/NASA)

What *COBE* couldn't do was map the temperature variations with high resolution. The *COBE* image had an effective resolution of about 10 degrees on the sky. Remember that the moon is about a degree on the sky, so the fluctuations were blurred to the size of 20 moons.

But the fact that there were fluctuations at all was very interesting to cosmologists. It meant that at a very early time in the universe (only about 300,000 years after the Big Bang), there were density variations in the universe at the level of 1 part in 100,000. It was quite possible that these density variations were the "seeds" that would eventually grow (thanks to gravity) into galaxies, galaxy clusters, and galaxy superclusters.

Cosmologists had often asked how, if the universe were indeed perfectly homogeneous, could stars, galaxies, or anything else form? Now they had a partial answer. The universe was not perfectly homogeneous. Something created density variations in the early universe. The CMB betrayed the presence of a little "crunchiness" in what had been thought to be a "smooth" early universe.

Boomerang: Nailing Down the Scale

The *BOOMERANG* (Balloon Observations Of Millimetric Extragalactic Radiation And Geophysics) mission flew for 10 days in late 1998 and early 1999. *BOOMERANG* did not try to compete with the *COBE* mission by mapping the whole sky, nor did it try to compete with NASA's upcoming *Microwave Anisotropy Probe* (MAP) mission by mapping the whole sky at high resolution. Instead, the balloon-borne craft flew 37 kilometers above Antartica, sampling high-resolution data at microwave frequencies, sensitive to the part of the spectrum where the universe emits photons that are a remnant echo of the Big Bang.

This image from the original BOOMERANG *data published in Spring of 2000 indicates the difference between the resolution of the* COBE *data and the* BOOMERANG *data. Notice that the* BOOMERANG *data covers only a small fraction of the sky, but at a much higher resolution.*

(Image courtesy of NASA/BOOMERANG)

BOOMERANG mapped only a tiny portion of the sky, but it did so with a much higher resolution than *COBE*. It made images at four frequencies, and all four frequencies were imaged with less than a degree resolution, more than 10 times "sharper" than the *COBE* images.

It's All About Geometry

The results were nothing short of breathtaking. *BOOMERANG* was able to detect the true scale size of the fluctuations in the cosmic microwave background. Are you on the edge of your seat yet? What did *BOOMERANG* find, you ask? It turns out that the scale size of the detected fluctuations was about 1 degree.

In a flat universe (one that has critical density, or $\Omega_0 = 1$) the scale of the fluctuations in the cosmic microwave background (CMB) is expected to be about 1 degree, or twice the size of the full moon. Fluctuations smaller or larger than this mean that the universe has a curvature. In a closed universe the fluctuations would be larger than 1 degree, and in an open universe, they would be smaller. This result implied a flat universe, a conclusion that reinforced the idea, set out in Chapter 20, that energy, not matter, must account for some 70 percent of the density of the universe.

The scale size of the temperature fluctuations in the microwave background can tell us about the geometry of the universe. A flat universe should have fluctuations with scale sizes of about 1 degree. In a closed universe the fluctuations would be larger, and in an open universe, they would be smaller. The BOOMERANG *data (center image) strongly suggested that we live in a universe with a flat geometry.*

(Image courtesy of NASA)

Hold On, I'm Going to Accelerate

While the *BOOMERANG* scientists were lending credence to the notion of a flat universe, other astronomers were busy observing very distant sources and making some surprising discoveries of their own concerning not the shape of the universe, but its age and its fate. As we mentioned in Chapter 19, many cosmologists are deeply interested in the correct value for the Hubble constant. This constant is basically the "slope" of the line describing how fast galaxies are receding from one another as a function of distance from one another. The reason for the interest is that the slope of this line can be used to tell us the age of the universe. And everybody wants to know how old the universe is, right?

But there is a more subtle effect as well. Ever since the Big Bang was proposed, astronomers have been curious about the fate of the universe. Will the universe expand forever, or will it eventually halt its expansion and collapse into another fireball?

One way to answer this question is to compare the expansion rate of the universe now (using galaxies close to us) to the expansion rate long ago (using galaxies far away from us). With good data, we should be able to figure out if the expansion of the universe is slowing down or not. And if it's slowing down at the right rate, then we know that we are in for a Big Crunch.

High-Redshift Type Ia Supernovae

This is exactly what two competing research groups—the Supernova Cosmology Project and the High-Z Supernova Search Team—started to do in the mid-1990s. The goal was to find a relatively common astronomical object that had a known luminosity—and therefore could be used as a "standard candle" to great distances.

The "standard candle" had to be very bright, so that it could be seen out to great distances. It turns out that Type Ia supernovae are the perfect objects. Type Ia supernovae are the explosions of white dwarfs that have been stealing material off of a binary companion. When the mass of the thieving black dwarf exceeds 1.4 solar masses, it will explode as a Type Ia supernova. These are 10 to 100 times brighter than a Type II supernova, which is the collapse and explosion of a normal high-mass star.

Type Ia supernovae had been studied long enough to have a known "light curve," or brightening and fading as a function of time, and most importantly, nearby supernovae of this kind seemed to have a very regular peak luminosity. So if one could just identify Type Ia supernovae (which can be done by analyzing their optical spectra), then their apparent brightness can be used to determine the distance to the galaxy in which the supernova was detected. The beauty is that a Type Ia supernova can be seen up to billions of light-years away.

Can This Be Right?

The two teams studied the data from a total of about 50 high-redshift Type Ia supernovae. They were surprised to find that the expansion of the universe was not slowing down at all. And what was truly shocking was that the expansion of the universe was, as noted in Chapter 17, speeding up. All the evidence from the high-redshift supernovae, which enabled researchers to trace the Hubble expansion back to the earliest times in the universe, indicated that the universe was accelerating in its expansion. It was expanding more rapidly now than it had in the past.

This was a result that no one had expected.

Astronomers understood an expansion that was slowing down as indicating that the universe had critical density, and that the force of gravity was winning out. But an *accelerating* expansion? What did that mean?

Was Einstein Right After All?

Remember how Einstein had to initially propose a "cosmological constant" for the universe? Without it, his theoretical universe always collapsed under its own gravity. The cosmological constant kept his universe pumped up. It provided a force that pushed things apart.

When Edwin Hubble discovered the recession of galaxies in the 1920s, the cosmological constant was put away, deemed unnecessary, and an abashed Einstein called it his "biggest blunder." Then in 1998, with the discovery of the accelerating expansion of the universe, the cosmological constant was pulled out of the drawer and dusted off. Could the force that it described (which followed naturally from Einstein's theory of General Relativity) be the force that was pushing things apart?

Now recall that despite their best efforts, astronomers had at the time been unable to account for more than about 30 percent of the critical density of the universe. Including luminous and dark matter, there is about 70 percent of the critical density entirely unaccounted for. Yet the evidence from the cosmic microwave background seemed to indicate that the universe had critical density, and that the universe had a flat geometry.

It is possible that the other 70 percent of the density of the universe is not in its mass, but in some form of energy. Remember that energy and mass are in some sense interchangeable (as Einstein famously described with the equation $E = mc^2$). In fact, the latest observations support about 25 percent dark matter and 75 percent dark energy, giving the universe its critical density.

Einstein's "blunder" might point the way. If the expansion of the universe is accelerating and the universe has "critical density," these might both be explained by energy contained in the vacuum of space itself.

Situation: Critical

Why should we expect that the universe has critical density? Does some human aesthetic bias compel us to envision a universe that has the density required to balance gravity exactly?

Not really.

Even before the *BOOMERANG* experiment found that the scale of the fluctuations in the CMB were about 1 degree (consistent with a flat universe and critical density), there were discussions among cosmologists that went something like this. The value of Ω (the density in mass and energy) could be anything. It is very unlikely that the value would be so close to 1 (remember observations of the universe indicate that dark and luminous matter make up 0.3), and not actually be 1.

But with *BOOMERANG* there was clear evidence that the universe had critical density. And more evidence has been coming in.

This Just In: The Whole Sky

The *BOOMERANG* mission did in some sense scoop the NASA *Microwave Anisotropy Probe* (MAP), now called the *Wilkinson MAP* (WMAP). *BOOMERANG* made the first determination of the scale of the fluctuations in the CMB. But the high-resolution image of the fluctuations in the CMB of the whole sky had to wait for the *WMAP*, which was launched in June 2001.

In February 2003, NASA and the *WMAP* team released the first high-resolution, all-sky images of the cosmic microwave background. These results have provided stunning confirmation of the *BOOMERANG* result and have placed even tighter constraints on the history of the early universe. Another important result from the *WMAP* is that the first stars appear to have formed when the universe was only about 200 million years old, much earlier than had been previously thought.

The Wilkinson Microwave Anisotropy Probe *(WMAP) has imaged the entire sky with less than 1 degree resolution, similar to the* BOOMER-ANG *image. This image shows the comparison between the* COBE *(top) and the* WMAP *(bottom) results.*

(Image courtesy of WMAP/NASA)

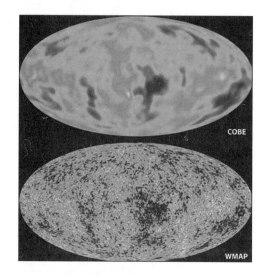

The Future? It'll Be a SNAP

The high-redshift supernova observations that began with ground-based observations in the 1990s have moved in part to observations with the *Hubble Space Telescope.* In the future, these research teams hope that the observations will be carried out by a dedicated satellite instrument, called *SNAP,* or the *SuperNova/Acceleration Probe.*

One of the limitations of the current high-redshift supernova work is that it is often difficult to get time on telescopes for follow-up observations on short notice. Follow-up observations are required to determine the type of a detected supernova and to track the evolution of the light curve. As a dedicated instrument, *SNAP* would resolve many of these observational issues and add to our list many more supernovae at very high redshifts.

Putting It All Together

Let's now pull together all the issues that have been clarified about the early history of the universe in the past five years.

◆ Based on the *WMAP* results and other recent observations, the universe has an age of 13.7 billion years, plus or minus only about 100 million years.

◆ The first stars in the universe started fusing hydrogen only 200 million years after the Big Bang, much earlier than previously thought.

♦ The universe consists of 4 percent normal atoms (the stuff we are made of), 23 percent dark matter, and 73 percent dark energy.

♦ The expansion rate of the universe gives us a value for the Hubble constant (H_0) of 71 km/sec/Mpc.

In 1920, Robert Frost wrote a poem called "Fire and Ice." "Some say the world will end in fire, / Some say in ice," Frost began. We might now have a definitive answer.

In the end, it looks as if we are just tiny part of a universe that is going to expand forever, at an ever increasing rate. The universe, then, will end in ice. Of course, that still leaves us with absolutely no idea what 96 percent of the universe is—right now, while it's still in business. In the final chapter we wonder if anyone else out there in the universe is pondering its content and its fate. And if so, will we ever be able to get together and talk about it?

The Least You Need to Know

♦ The cosmic microwave background (CMB), which is our insight into the earliest time of the universe, indicates that the universe is not perfectly homogeneous; tiny fluctuations in density were present even when the universe was in its infancy, only 300,000 years old. It was first explored in detail by the *Cosmic Microwave Background Explorer (COBE)*.

♦ The scale of the fluctuations in the CMB, imaged first by the *BOOMERANG* mission and later by NASA's *Wilkinson Microwave Anisotropy Probe (WMAP)*, is about 1 degree: the size of fluctuations expected in a "flat" universe.

♦ Astronomers in two research groups are using Type Ia supernovae to determine exactly how the universe expanded in its early years. The expansion of the universe appears to be accelerating.

♦ Combining the results of high-redshift supernovae and the fluctuations in the cosmic microwave background suggests that most of the density of the universe is not in visible matter or dark matter, but in dark energy, whatever that might be.

♦ It looks like the universe is about 14 billion years old, has a flat geometry, and consists of about 4 percent atoms, 23 percent dark matter, and 73 percent dark energy. That's a nice way of saying that we don't know what 96 percent of the universe is. We'd better get to work.

Chapter 22

Where Is Everybody?

In This Chapter

- ◆ The Fermi Paradox
- ◆ What is life?
- ◆ Is the earth unique or typical in its ability to sustain life?
- ◆ Is there other life in our solar system?
- ◆ The case for life on Mars
- ◆ Life in our galaxy: the Drake Equation

Enrico Fermi posed a question in the 1950s that has since become known as the Fermi Paradox: If life is abundant in the universe, and the universe is both vast and old, where is everybody? The question goes to the heart of something that humans have wondered for a long time. Are we alone?

There are a number of ways to consider resolving the paradox. Is the universe perhaps teeming with life that can't get to us (because of the limitations and cost of space travel), or are we the first wave of intelligent life in the Milky Way, or are others so advanced as to be uninterested in us? This chapter confronts some of these questions, and will help you consider your own answers to this important question.

In the past four centuries, we have learned that there is nothing special about our position in the cosmos: We inhabit the third planet orbiting an average star in a solar system tens of thousands of light-years from the center of a normal spiral galaxy. And there are a few hundred billion other stars in the Galaxy. Granted, not all of them could have solar systems like our own, but the history of our understanding of our place in the universe should tell us that an assumption of mediocrity is probably a good idea.

And if we're not special, why has no one contacted us? Do we smell bad? Is it our hair? Or is our planet truly the only repository of life in the universe? Let's take an inventory of what we know, and what it might tell us about our place in the universe as intelligent, curious beings.

What Do You Mean by "Alone"?

Sitting in your home at night, watching television, with no other humans present, you might consider yourself to be alone. But are you really alone? There are microbes crawling on your skin, bacteria in your intestines, your cat sprawled across your lap. Silverfish inhabit the damp cabinet underneath your kitchen sink. And yet you might still consider yourself "alone." Why is that?

Although there are other living things in your proximity, some even on or in your own body, you cannot communicate with them. Your cat comes close, but even she has limitations beyond appreciating being fed and scratched.

Now, reconsider the larger question. Is the mere existence of other life in the universe enough for us not to feel alone? Would there be any comfort in knowing that the universe is teeming with bacterial life? Or will our cosmic loneliness be ended only by the presence of other intelligent life, life we could potentially communicate with?

... If You Call This Living

So in thinking about the prospects for life beyond our planet, most of us probably contemplate two distinct subjects: intelligent life—life capable of creating civilizations—and, well, just plain life.

How plain can life get and still be called life?

You know how a shrink-wrapped box of software has "minimum requirements" printed on its side? How many megabytes of RAM you need, how much hard drive space, and so on? Well, life seems to have some minimum requirements as well. Most biologists would agree that the requirements for something to be classified as a living organism include ...

◆ The ability to reproduce.

◆ The ability to obtain nourishment or energy from the environment.

◆ The ability to evolve or change as the environment changes.

Biologist Steven J. Gould has called the current age on Earth the Age of Bacteria, and he makes the case that, in terms of diversity and adaptability, "bacteria rule." Kurt Vonnegut Jr. once suggested that the purpose of humans is to bring viruses to other planets.

Bacteria and viruses might be plentiful, but what about more macroscopic beings? Could these be found—or even be common—on other planets?

Is Earth Rare?

The requirements listed previously are minimal and, for example, say nothing about movement, specific senses, or even consciousness (more specific attributes such as these narrow and refine the definition of life). We know that life on Earth ranges from tiny viruses (which hardly seem alive at all when inactive, and which many biologists do not classify as living) to simple one-celled organisms to the whole range of more complex plant and animal life to ourselves.

Life at its most basic seems to require very little: the presence of a few chemical elements, the absence of certain harmful substances, and the availability of tolerable environmental conditions. Some scientists believe that the existence of these basic requirements on Earth is a sort of cosmic fluke—a rare, perhaps unique, lottery jackpot win; but most assume that Earth, although special to us, is (in the Galaxy) mediocre. The belief that there is nothing uncommon, let alone unique, about the conditions on Earth that support life is sometimes called the *assumption of mediocrity*.

Star Words

The **assumption of mediocrity** basically states that we are not special in any way, and that as part of that assumption, life on Earth is not unique. The conditions that have allowed life to arise and evolve on Earth must exist many other places in the Galaxy and universe.

As far as we can tell, Earth does not possess any special elements or conditions that aren't easily available elsewhere in the Galaxy and the universe. Not only, then, is there nothing to have prevented life from evolving elsewhere in the universe, but with hundreds of billions of stars in a single galaxy, there is every reason to believe that life *has* indeed sprung forth elsewhere.

There are even recent suggestions that life might have originated on Mars and was brought to Earth on the backs of asteroids. Other researchers have been able to generate structures that could be the precursors to cell walls in conditions that simulate the harsh environments of interstellar space. Life, once thought to be fragile and unique, appears to be a tough survivor.

But Fermi's question remains, "Where is everybody?"

The Chemistry of Life

According to those who study high-redshift supernovae, the universe is about 14.5 billion years old. Earth, like other planetary bodies in the solar system, appears to be about 4.5 billion years old, and the fossil record shows that Earth was not devoid of life for very long, perhaps for only a few hundreds of millions of years after it coalesced from the solar nebula. The oldest fossils (Precambrian microfossils) known on Earth are from Western Australia, and are about 3.5 billion years old. They resemble living bacteria. The first fossils indicating the presence of life requiring free oxygen are about 2 billion years old, which means that Earth's atmosphere underwent a huge change in the intervening billion and a half years. The fossil record tells us that life arose relatively quickly on the surface of Earth, and went through waves of diversity and evolution on timescales of tens of millions of years. One glaring mystery remains: How did life get a foothold in the first place?

Most scientists who study life on early Earth propose that some time before these first life forms arose, during Earth's first several hundred million years, the simple molecules present (nitrogen, oxygen, carbon dioxide gas, ammonia, and methane) somehow formed into the more complex amino acids that are the chemical building blocks of life. The turbulent youth of Earth provided the energy that caused the transformation of simple elements and compounds into the building blocks of life. Also remember that, in its infancy, Earth was orbited by a much closer Moon, so that after liquid water was present on its surface, the tides would have been enormous—perhaps 1,000 feet up and down as the planet rotated once every 6 (not 24) hours! This vigorous mixing of the primordial soup (shaken, not stirred) might have had an important effect on the origin of life, providing a "workbench" on which chemical experiments could occur.

Astro Byte

Volcanic activity on the early Earth sent many gases coursing into the atmosphere. Its early carbon dioxide atmosphere literally arose from within. Earth had sufficient mass to hold onto the heavier molecules in its atmosphere, but not the lighter hydrogen and helium. Earth was a warmer place in the past, warmed by its greenhouse gases.

After amino acids and nucleotide bases were available (and observational evidence of star-forming regions suggests that highly complex molecules are present in interstellar space), the next step up in complexity would have been the synthesis of proteins and genetic material. The DNA molecule is made up of nucleotide bases called genes. Strung together, the genes tell our cells what to do when they reproduce, and they make us different (in some ways) from, say, earthworms. The DNA molecule is the most durable, portable, and compact information storage device we know of, and we are just beginning to unravel its mysteries.

But this is not to say that DNA is the only way that life could propagate. Perhaps there are myriad means by which life can encode ways to make more of itself. Nevertheless, a starting point is to understand the genetic material that supports life on Earth, and we are in the early phases of that understanding.

Although there might be plenty of planets similar to Earth, it is extremely difficult to estimate the likelihood that the molecules on any given planet will combine to form amino acids and nucleotides, let alone proteins and DNA. *Astrobiologists* have wondered: Are there fundamental biological laws in the same way that there are fundamental physical laws? Do the basic, universal rules of chemical equations inevitably lead to information storage systems like DNA, or is DNA a fluke peculiar to our planet? Here is where the questions, and the debates, begin.

Star Words

Astrobiologists are scientists who are able to contemplate questions (like the origin and evolution of life on Earth) that arise at the intersection of these two fields. Astrobiology is a relatively new discipline, combining the expertise of astronomers and biologists.

Close Encounter

In 1953, scientists Stanley Miller and Harold Urey decided to see if they could duplicate, experimentally, the chemical and atmospheric conditions that produced life on Earth. In a 5-liter flask, they replicated what is thought to be Earth's primordial atmosphere: methane, carbon dioxide, ammonia, and water. They wired the flask for an electric discharge, the spark intended to simulate a source of ultraviolet photons (lightning) or other form of energy (such as a meteor impact shock wave—which must have been common in the early solar system). The Miller-Urey Experiment didn't produce life, but it did create a collection of amino acids (building blocks of proteins), sugars, and other organic compounds. Thus the Miller-Urey Experiment implied an extension of the assumption of mediocrity. Not only could we reasonably conclude that there must be similar planets in the Galaxy, but also that they must have the chemical elements necessary for life, and random energy alone could trigger the synthesis of the building blocks of life.

All life on Earth is carbon-based. That is, its constituent chemical compounds are built on carbon combinations and, furthermore, developed in a liquid water environment. Even creatures such as ourselves, who do not live in water, consist mostly of water, and the presence of liquid water is vital to our continued existence. If life on Earth developed first in the oceans, we, billions of years later, individually develop as fetuses in a tiny oceanlike womb, and then we—being mostly water—carry these oceans within us through our lives.

The Odds for Life on Mars

Science fiction has long portrayed Mars as home to intelligent life, and Percival Lowell, early in the twentieth century, created a great stir with his theory of Martian *canals*. Actor-director-playwright Orson Welles triggered a nationwide panic with his 1938 radio dramatization of H. G. Wells's *War of the Worlds*, about a Martian invasion of Earth. People of Earth seem primed to believe that life might exist on Mars.

Given our fascination with the red planet and its proximity, it is no wonder that Mars has been the target of a number of unmanned probes. In the mid- to late 1970s, the *Viking* probes performed robotic experiments on the Martian surface, including tests designed to detect the presence of simple life forms such as microbes. These tests yielded apparently positive results, which, however, were subsequently reinterpreted as false positives resulting from chemical reactions with the Martian soil. Then, on August 7, 1996, scientists announced that, based on its chemical composition, a meteorite recovered in Antarctica had originated on Mars and possibly contained evidence of fossilized bacterial life.

The two Mars Exploration Rovers *will be able to function as robotic scientists, extending the reach of human understanding.*

(Image from NASA)

Most recently, the *Mars Pathfinder* (1997) and *Mars Global Surveyor* (1998) missions have added detailed panoramic views from the Martian surface and high-resolution satellite images of the surface. Both of these missions suggest the presence at some time in the past of liquid water on the Martian surface. In January 2004, two Martian rovers launched in the summer of 2003 by NASA touched down on the surface of Mars, in search of information about the history of water on Mars. For more on what was found, see Chapter 8.

All of these results are intriguing, but, so far, none of the Martian probes has found clear evidence of life (past or present) on Mars. In the past, the Martian atmosphere, now very thin, was likely thicker and, as a consequence, the planet's surface warmer and wetter. There is a possibility that microbial life once existed on Mars. Under the planet's current cold, dry, and generally harsh conditions, however, the presence of life is highly unlikely.

The Face on Mars

In 1976, *Viking 1*, surveying Martian landing spots for *Viking 2*, took a picture of the Martian surface that revealed what some observers interpreted to be a giant monument, about a third of a mile (0.5 km) across, in the form of a human (or humanoid?) face. The Face on Mars has provoked much discussion in the popular press and (more recently) on the Internet. With the canals on Mars long since revealed as nonexistent, the Face on Mars provided new hope, as some saw it, that intelligent life either exists or once existed on Mars. Although scientists never took the Face on Mars seriously, some popular commentators even suggested that objects resembling pyramids and other structures could be seen in the vicinity of the face.

Close Encounter

Europa (see Chapter 10), a moon of Jupiter, has a frozen surface that might have liquid water beneath it. Titan (see Chapter 10), orbiting Saturn, has an atmosphere rich in nitrogen, methane, and ammonia. Titan will be observed by a probe as part of the *Cassini* mission to Saturn that will reach the planet in early 2004. Titan might also have liquid methane on its surface. So far, these planetary moons, rather than any planet, offer the best prospects for other life within the solar system. However, the prospects are dim, as surface temperatures on both of these moons are extremely low.

On April 5, 1998, the Mars Orbiter Camera on the *Mars Global Surveyor* made high-resolution images of the "face," showing it to be nothing more than a natural formation, much like mesas found in the American Southwest. The angle of sunlight and the comparatively low resolution of the 1976 image, combined with the ever-fertile human imagination, had temporarily transformed this geological feature into a giant face.

The Face on Mars—as photographed by Viking 1 *on July 25, 1976 (left), and as photographed on April 5, 1998, by the* Mars Orbiter Camera *on the* Mars Global Surveyor *(two images on the right).*

(Images from NASA/MSS)

Hello! Is *Anybody* Out There?

Life beyond Earth but within the solar system seems unlikely. But that's okay. There's still the rest of the galaxy, right?

With hundreds of billions of stars in our own galaxy, odds are that there must be many other planetary systems. In fact, observations of the "wobble" of stars (see Chapter 11) indicates the presence of well over 100 stars with planets around them. Recent observations have shown that stars with high metallicity (determined by the relative abundance of iron in the star) are much more likely to have planets. But that narrows the odds only slightly. And if one of these planets orbits at a constant distance from its host star at a distance at which liquid water could exist, couldn't life exist there as well? How do we determine the odds?

An Equation You'll Like

Given the billions of stars in our galaxy, the existence of life somewhere in the Milky Way is pretty likely, but is there any way to get a rough idea of just how likely? Not every star in the Galaxy is a good prospect for life. Some stars are so massive that they don't last long enough, and others are not hot enough to warm any planets that might be close by. Others formed long ago when there were not as many metals present, and others are in binary systems (in which simulations show planets have a harder time surviving).

Perhaps 15 percent of the stars in the Galaxy have the proper mass to be luminous enough (but not too luminous) to support habitable planets. In 1961, an astronomer named Frank Drake attempted to quantify the odds that intelligent life capable of communication exists elsewhere in the Galaxy. He tried to get at least a rough answer to part of the Fermi Paradox.

Drake proposed an equation to estimate the possible number of civilizations in the Milky Way. The *Drake Equation* contains a number of highly uncertain terms—variables that must simply be guessed at—however, it is a very useful way of breaking down a complex question into smaller, differentiated blocks. Some of the variables that were highly uncertain in 1961—the number of stars with planets, for example—have become more certain as further relevant observations were made.

Star Words

The **Drake Equation** consists of eight terms that help to make a rough estimate of the number of communicating, intelligent civilizations in our Galaxy, the Milky Way.

Here's what the Drake Equation looks like:

$$N = R_* \times f_p \times N_p \times f_e \times f_l \times f_i \times f_c \times L$$

And here's what the terms of the Drake Equation stand for:

- N (the left-hand side of the equation) is the number of civilizations in our Galaxy with which we should be able to communicate via some sort of signal.

- R_* is the rate at which our galaxy forms solar-mass stars—its productivity in solar masses per year.

- f_p is the fraction of stars with planetary systems.

- N_p is the average number of planets per star.

- f_e is the fraction of Earthlike planets (planets suitable for life).

- f_l is the fraction of these planets on which life actually develops.

- f_i is the fraction of these planets on which intelligent civilizations arise.

- f_c is the fraction of these planets on which technological civilizations develop.

- L is the lifetime of a civilization in years.

A Closer Look at the Equation

Perhaps the most important term in the equation is L, the lifetime of a civilization. If civilizations last for many millions of years, then our own Galaxy could be teeming not only with life, but also with advanced civilizations. As Fermi and others realized, the universe has been around for a long time. The types of stars around which civilizations could evolve have been present since relatively soon after the Big Bang, and our planet took about 5 billion years to produce us. But Earth didn't form until the universe was almost 10 billion years old. If the likelihood of all of these variables is high, intelligent civilizations should have colonized the Galaxy by now. So where are they?

Here's one somewhat depressing possibility: If L is rather small (hundreds or thousands of years), then the Galaxy might be a lonely place, indeed. Civilizations, like candles, might be lit frequently, but if they do not burn for very long, only a few others (or no others) will be burning with us.

Let's look at some of the other key terms in the equation.

Galaxy Productivity

Astronomers figure that a minimum of 100 billion stars populate the Milky Way. We can estimate that about 10 percent of these stars (or about 10 billion stars) are solar-mass stars that are not in binary systems. If the Galaxy is about 10 billion years old, then the star formation rate is about 10/10, or 1 solar-mass star each year.

Do They All Have Planets?

When the Drake Equation was first written down, the only planetary system we knew about was our own. But astronomers have expended great effort in the past three decades to uncover other planetary systems (see Chapter 11). As noted previously, we now know that there are many other planetary systems, over 100 at last count, and a recent study based on statistical corrections to planetary searches estimates that at least 25 percent of solar-mass stars have planetary systems.

Welcome to the Habitable Zone

Estimates (based on numerical modeling) of the typical number of planets per planetary system range from 5 to 20. The final number is highly dependent on the simulation that is run, and that shouldn't be surprising. There is some degree of randomness in the formation of a planetary system.

Our own solar system, of course, has nine planets. We have observations of other planetary systems, but we must remember that we can at present detect only the most massive planets orbiting these stars. We have no counts of the number of smaller planets in these systems.

However, many planets do form around a star; there is a doughnut-shaped region around all stars that astronomers call the *habitable zone*. This zone is the space between some inner and outer distance from the star where a star with a given luminosity (rate of energy production) produces a temperature that's "just right." Just right for what? Well, just right for liquid water to exist without freezing or boiling away. We will estimate that 10 percent of the planets around a star will fall within the habitable zone. Recent work has narrowed this zone by raising the concept of a "continuously habitable zone." This is the distance from a star (throughout its changeable lifetime) at which liquid water could exist. It is interesting to note that although Venus, Earth, and Mars fall in the habitable zone of the Sun, only Earth falls in the (narrower) continuously habitable zone.

> **Astronomer's Notebook**
>
> Three planets are in the **habitable zone** of the Sun—Venus, Earth, and Mars. The habitable zone is the distance from the star at which liquid water could survive. So apparently being in the "zone" is not sufficient to support life!

Primordial Soup du Jour

Although astronomers such as Carl Sagan (1934–1997) have assumed that life develops on virtually every planet capable of supporting it, estimates vary widely. Until scientists produce a model whereby basic elements can self-organize into reproducing "proto-life," the odds of life arising given the right conditions are anybody's guess.

The fraction of habitable planets on which life arises, *fl*, is expressed as a decimal, with the number 1 indicating that there is a 100 percent chance that life will develop on a habitable planet. Estimates range from Sagan's optimistic 1 down to 0.000001— that is, perhaps only 1 out of every million habitable planets actually develops life. This number is highly uncertain. If robotic probes in the solar system were to find that microbial life once existed on Mars, then this might cause us to raise the odds that life will arise given the right conditions.

As you will see later, the probability that simple life arises might be high, only to be undone later in the equation by the difficulties presented to long-term survival of complex life.

Intelligent Life? Where?

Beyond this point in the equation, the terms become very uncertain. Some scientists believe that it is possible that biological evolution as observed on Earth is a universal

Kurt Vonnegut once said that the ultimate purpose of humans might be to bring viruses to other planets. The survival of unprotected microbes on the metal surfaces of landing craft on the Moon between two Apollo missions indicates that some simple life forms are indeed much hardier than we ever expected.

phenomenon; given enough time, life anywhere it exists will tend to evolve toward greater and greater complexity.

If this is the case, then we might assume that intelligence will emerge (given sufficient time) wherever there is life. Others, however, believe that intelligence is not indispensable to survival and that, therefore, it is not the inevitable byproduct of natural selection and evolution. Are bacteria intelligent? Not particularly, and they've done pretty well for themselves here on Earth.

Turn on the Radio

For the fraction of intelligent life forms that develop civilization and technology (f_c), estimates have a wide variation. How could they not? We have now entered the realm of psychology and social science, disciplines that have a hard enough time understanding humans, much less other life forms.

Some argue that, given time, technology and civilization are inevitable developments of intelligence. Others make just as compelling a case that the connection is not inevitable, and they consequently put this figure much lower. Other complicating factors involve the correlation between intelligence and aggression. Are intelligent life forms destined to destroy themselves? Or do a few select civilizations enter a time of peace, a *pax planetaria?*

Finally, it cannot be assumed that all civilizations will have the *desire* to communicate. Maybe civilizations develop virtual realities much more interesting than the real universe, and develop beyond physical, biological bodies all together. This term in the Drake Equation has been rich territory for mining by good science fiction writers.

It's the End of the World As We Know It

The final term is perhaps both the most interesting and the most uncertain of them all: How long do civilizations capable of communication last? To estimate how many civilizations there are in the Galaxy at any given time, we must know how long they last. Once lit, how long does the candle of civilization (on average) burn? Consider our own civilization, one with a recorded history of no more than 5,500 years and one that has been highly technological (possessing the atomic bomb and radio telescopes) for less than a century.

Planets and stars have long life spans, but we have only a little over 5,000 years of documented experience with civilization. What is most disquieting is that we have reached a point of technological development that enables us, at the touch of a few buttons, to destroy our civilization in a thermonuclear fireball here and now. And the solar system is still a place where things are out of our control. Were an asteroid like the one that struck Earth about 65 million years ago to hit us again, we would fare no better than the dinosaurs. Fifty million years after that impact, highly evolved ants might be excavating our fossilized bones and speculating on the cause of our demise.

So what is the average life span of a civilization? A thousand, five thousand, ten thousand, a million years? Even if, over time, there have been many, many civilizations in our galaxy, how many are there now?

Were an asteroid like the one that doomed the dinosaurs to hit Earth tomorrow, we would fare no better than they did. Civilizations face internal and external challenges to their survival.

(Image from NASA)

Close Encounter

If we approach any variable in the Drake Equation pessimistically, assigning it a low value, we end up with the possibility that there are very few technological civilizations in the Galaxy. But if we plug into the equation the most optimistic estimates for all the variables (such as life will always develop, intelligence will always evolve, technology will always arise), we end up finding that the number of technological civilizations in the Galaxy is equal to the average lifetime of such a civilization.

If the average technological civilization survives 10,000 years, we can expect to find 10,000 civilizations in the Galaxy at any time. If 5,000 years is the norm, 5,000 civilizations would be expected. If a civilization usually lasts millions of years, then, assuming high values throughout the rest of the equation, we might expect to find a million civilizations in the Milky Way. But even in this optimistic scenario, the distances between civilizations remain enormous.

Traveling to the location of another civilization in the Galaxy is a highly impractical proposition. Even in a best-case scenario, civilizations in the Galaxy must be hundreds of light-years apart, and getting to the nearest star would take us (using current technology) tens of thousands of years. A civilization might be long gone by the time we got there—to say nothing of the mother of all jet lags incurred by the voyagers. Wouldn't it be tragic to travel to another star system to find that they had done themselves in during the thousands of years it took us to get there?

Again, I Say, Where Is Everybody?

The original goal of the Drake Equation was to justify that the search for life in the Galaxy was not a hopeless venture, that the Galaxy might indeed be teeming with life, and that if we just looked, we might well find something or someone looking for *us*.

What We Look For

Given the vast distances that must separate civilizations even in the most optimistic scenarios, it is unlikely that we will be traveling to distant civilizations any time soon.

But, the urge to detect some sign of other intelligent life in the universe persists, and since the 1960s, dedicated groups of researchers have trained radio telescopes skyward in the hope of receiving broadcasts from extraterrestrial sources. Searches have varied from targeted (directed toward other Sunlike stars) to random (piggybacking on other astronomical observations). So far, none of the searches have detected a repeating signal of extraterrestrial origin.

Astro Byte

Why radio signals? Why not an optical laser pulse sent out into space? It is certainly possible, but because the dust in the plane of our Galaxy absorbs optical photons very well, the energy requirements for optical communication would be much greater. Radio waves, on the other hand, pass through the plane of the Galaxy as if it weren't there. We (and our potential neighbors) can cast a cheaper, wider net in radio waves.

"Earlier on *Survivor* ..."

The project of monitoring the heavens for radio broadcasts seems to assume that a technological civilization would have made the decision intentionally to broadcast a repeating signal. Given that we earthlings are doing no such thing, this assumption seems a bit shaky.

There is no need to assume, however, that the civilization actually *wants* to communicate and is intentionally broadcasting to other worlds. After all, we on Earth have been broadcasting radio waves since the early twentieth century and have been doing so intensively for more than 60 years.

Although radio signals at longer wavelengths do not penetrate beyond our atmosphere, those in the higher frequencies, FM radio and television, are emitted into space. We don't *intend* this to happen, but it is an inevitable by-product of our technological civilization (at least until everything comes through optical fibers). Perhaps some other civilization is producing a similar by-product. Our first glimpse of another civilization might be the equivalent of its sitcoms, advertising, reality TV, and televangelists. Yet we keep on listening.

The SETI Search

Monitoring the heavens for artificial extraterrestrial broadcasts is a daunting task—akin to looking for a needle in a haystack. Where should we look? What frequencies should we monitor? How strong will the signal be? Will it be continuous or intermittent? Will it drift? Will it change frequency? Will it even be recognizable to us?

To say the least, radio frequency searches are time- and equipment-intensive. Frank Drake conducted the first search in 1960, using the 85-foot antenna at the National Radio Astronomy Observatory in Green Bank, West Virginia. He called his endeavor Project Ozma, after the queen of the land of Oz. This search ultimately developed into Project SETI (Search for Extra-Terrestrial Intelligence), which, despite losing Federal funding in the 1990s, remains the most important sponsor of search efforts.

The SETI Institute is a private, non-profit group based in Mountainview, California. When NASA-based SETI funding was cut in 1993, SETI consolidated much of its effort into Project Phoenix (risen, like the mythical bird, from the ashes of the funding cut), a program that began in 1995 and that monitors 28 million channels simultaneously. SETI hopes eventually to monitor two billion channels for each of some 1,000 nearby stars. Computer software alerts astronomers to any unusual, repeating signals. So far, no repeating signals have been discovered.

Close Encounter

Want to do your bit for the Search for Extra-Terrestrial Intelligence (SETI)? Download a screensaver (at www.seti-inst.edu/science/setiathome.html) that might help to find that first ET. The program uses cycles on your computer (when not in use) to process one of the largest datasets mankind has ever assembled.

And if you just want to check the current status of SETI, go to their informative website: www.seti-inst.edu. The SETI Institute recently received a large financial boost from Microsoft cofounders Paul Allen and Nathan Myhrvold, who have committed $25 million to the construction of a radio telescope array dedicated to the SETI search. Up to this point, searches have depended on cadging time on existing telescopes. The proposed Allen Telescope Array will be dedicated to the SETI project around the clock.

Down at the Old Water Hole

SETI researchers also guess that civilizations might choose intentionally to broadcast their presence to their neighbors, sending some sort of radio-frequency beacon into space. What portion of the spectrum might they choose?

Researchers chiefly monitor a small portion of the radio wavelengths between 18 cm and 21 cm, called the *water hole.* The rationale for monitoring this slice of the radio spectrum is twofold. First, the most basic substance in the universe, hydrogen, radiates at a wavelength of 21 cm. Hydroxyl, the simple molecular combination of hydrogen and oxygen, radiates at 18 cm. Combine hydrogen and hydroxyl, and you get water.

If the symbolism of the water hole is not sufficiently persuasive to prompt extraterrestrial broadcasters to use these wavelengths, there is also the likelihood that this slice of the spectrum will be recognized as inviting on a more practical level. It is an especially quiet part of the radio spectrum. There is little interference here, a very low level of galactic "background" noise. Researchers reason that, at the very least, intentional broadcasters would see the water hole (which looks the same no matter where in the Galaxy you are located) as a most opportune broadcast channel.

Do We Really Want to Do This?

Hollywood has done a pretty fair job of reflecting the range of opinion on the psychological makeup of extraterrestrial beings. The 1950s and even 1960s saw a number of movies about malevolent alien invaders; but then Steven Spielberg's 1977 *Close Encounters of the Third Kind* and *E.T.—the Extraterrestrial* (1982) suggested that contact with emissaries from other worlds might be thrilling, beneficial, and even heartwarming.

The close of the 1990s pitted *Independence Day*—in which aliens attempt to take over Earth—against *Contact*, in which communion with extraterrestrials is portrayed as a profoundly spiritual (if excessively teary-eyed) experience. And then *X-Files: Fight the Future* proposed that our entire planet is being prepared for massive colonization by extraterrestrials. In fact, the whole *X-Files* television series suggested a novel and surprisingly popular solution to the Fermi Paradox. Where is everybody? Hey, they're already here!

Astro Byte

The *Voyager* spacecraft each have onboard a golden plaque and phonograph record. Among other things, the plaque shows a man and woman, the trajectory of the spacecraft in the solar system, and a representation of the hydrogen atom. The phonograph record contains recordings of human voices and music. It is a message in a bottle cast into a vast ocean.

The point is this: Not everyone is persuaded that reaching out is such a great idea. After all, even the most sociable among us lock our houses at night.

But, some argue, the point is moot; the radio cat has been out of the bag for more than six decades. There is an ever-expanding sphere of our radio broadcasts moving out at the speed of light in all directions from Earth. The earliest television broadcast (which, unfortunately, includes images of Adolf Hitler at the 1936 Olympics) is now almost 70 light-years from Earth. It's a bit late to start feeling shy.

On November 16, 1974, the giant radio dish of the Arecibo Observatory (see Chapter 5) was used as a transmitter to broadcast a binary-coded message containing a compact treasure trove of information about humans. Reassembled, the message generates an image that shows the numbers 1 through 10; the atomic numbers for hydrogen, carbon, nitrogen, oxygen, and phosphorous; a representation of the double-helix of DNA; an iconic image of a human being; the population of the Earth; the position of Earth in the solar system; and a schematic representation of the Arecibo telescope itself. No reply has yet been received. That should come as no surprise. Traveling at the speed of light, the message will take 25,000 years to reach its intended target, a globular cluster in the constellation Hercules. Maybe by then we will know how to greet our alien friends hospitably.

The Arecibo message was broadcast from the Arecibo radio telescope in Puerto Rico in 1974. The message is 1679 bits long, the product of two prime numbers, 23 and 73. When the zeros and ones are arranged in a rectangle with sides 23 and 73 units long, they make this picture, which shows among other things a human form, our number system, and the double helix of DNA.

(Image from NASA)

Some might view the Arecibo Message as the work of latter-day Quixotes jousting with cosmic windmills. Others might view it as an interstellar message cast adrift in a digital bottle, containing the very human hope of contact. Let's just hope that we haven't told them too much.

Coming Full Circle

We've come a long way since we considered our astronomical ancestors, looking up at the sky and wondering. We have come to the end of our story, a story that—as is obvious from the daily paper—is still being written. We live in an incredible age, when the discovery of accretion disks around black holes and distant Type Ia supernovae at the edge of the visible universe are reported on the evening news, along with doings of movie stars and musicians.

The Least You Need to Know

- The chemicals and conditions on Earth that support life are probably not unique or even exceptional, but common throughout the Galaxy and the universe.

- Some of the basic building blocks of life can be produced in the laboratory, and even have been detected in interstellar space.

- Although the possibility of life beyond Earth within our solar system is very small, the existence of simple life elsewhere in our Galaxy is highly likely.

- The Drake Equation is a rough way to calculate in a reasoned manner the number of civilizations that might exist in the Milky Way. It includes eight factors, some based in astronomy and well-determined, and others that are still highly speculative.

- Small but dedicated groups of researchers (including those of the SETI project) monitor the heavens for artificial radio signals of extraterrestrial origin. The SETI project is now on a very firm financial footing and is planning to build a dedicated SETI telescope.

Star Words Glossary

absolute magnitude   *See* luminosity.

accelerating expansion   Recent observations of very distant supernovae suggest that the expansion of the universe is accelerating; that is, the universe is expanding more quickly now than it did in the distant past.

accretion   The gradual accumulation of mass; usually refers to the build-up of larger masses from smaller ones through mutual gravitational attraction.

active galaxy   A galaxy that has a more luminous nucleus than most galaxies.

altazimuth coordinates   Altitude (angular distance above the horizon) and azimuth (compass direction expressed in angular measure from due north).

altitude   *See* altazimuth coordinates.

angstrom   Abbreviated A (or Å). A unit used to measure very small size, equal to one ten billionth of a meter, or 10^{-10} meter.

angular momentum   The rotating version of linear momentum. Depends on the mass distribution and rotational velocity of an object.

angular separation   Distance between objects in the sky expressed as an angle (such as degrees) rather than in distance units (such as feet or meters).

angular size   Size expressed as an angle (such as degrees) rather than in distance units (such as feet or meters).

annihilate   Used as an intransitive verb by astronomers. Particles and antiparticles "annihilate" when they meet, converting their mass into energetic photons. Electrons and anti-electrons (positrons) were continually created and annihilated in the early universe.

anthropic principle   Comes in different varieties, but the basic concept is that the universe must be the way it is (in terms of its fundamental parameters) or human beings would not be here to observe it.

apollo asteroids   Asteroids with sufficiently eccentric orbits to cross paths with Earth (and other terrestrial planets).

apparent magnitude   A value that depends on the distance to an object. *See also* luminosity.

arcminute   One sixtieth of an angular degree.

arcsecond   One sixtieth of an arcminute ($\frac{1}{3,600}$ of a degree).

assumption of mediocrity   A scientific assumption that we on Earth are not so special. The conditions that have enabled life to arise and evolve on Earth likely exist in many other places in the Galaxy and universe.

asterism   An arbitrary grouping of stars, within or associated with a constellation, perceived to have a recognizable shape (such as the Teapot or Orion's Belt) that readily serves as a celestial landmark.

asteroid   One of thousands of small, rocky members of the solar system that orbit the Sun. The largest asteroids are sometimes called minor planets.

astronomical unit (A.U.)   A conventional unit of measurement equivalent to the average distance from Earth to the Sun (149,603,500 kilometers, or 92,754,170 miles).

autumnal equinox   The date (usually approximately September 21) on which day and night are of equal length because the Sun's apparent course against the background stars (the ecliptic) intersects the celestial equator.

barred-spiral galaxy   A spiral galaxy that has a linear feature, or bar of stars, running through the galaxy's center. The bar lies in the plane of the spiral galaxy's disk, and the spiral arms typically start at the end of the bar.

Big Bang   The primordial explosion of a highly compact universe; the origin of the expansion of the universe.

binaries   Also called binary stars. Two-star systems in which the stars orbit a common center of mass. The way the companion stars move can tell astronomers much about the individual stars, including their masses.

black body   An idealized (theoretical) object that absorbs all radiation that falls on it and perfectly re-emits all radiation it absorbs. The spectrum (or intensity of light as a function of wavelength) that such an object emits is an idealized mathematical construct called a black-body curve, which can serve as an index to measure the temperature of a real object. Some astronomical sources (such as stars) can be approximated as black bodies.

black hole   A stellar-mass black hole is the end result of the core collapse of a high-mass (greater than 10 solar mass) star. It is an object from which no light can escape within a certain distance (*see* Schwarzschild radius). Although space behaves strangely very close to a black hole, at astronomical distances, the black hole's only effect is gravitational.

Bode's Law   Also called the Titius-Bode Law, Bode's Law is a numerical trick that gives the approximate interval between some of the planetary orbits in our solar system.

brightness   The measured intensity of radiation from an object. The brightness of astronomical objects falls off with the square of the distance.

brown dwarf   A failed star; that is, a star in which the forces of heat and gravity reached equilibrium before the core temperature rose sufficiently to trigger nuclear fusion.

calderas   Craters produced not by meteoroid impact but by volcanic activity. *See also* corona.

cardinal points   The directions of due north, south, east, and west.

Cassini division   A dark gap between rings A and B of Saturn. It is named for its discoverer, Gian Domenico Cassini (1625–1712), namesake of the mission that will arrive at Saturn in 2004.

cataclysmic variable   *See* variable star.

celestial equator   An imaginary great circle dividing the Northern and Southern Hemispheres of the celestial sphere.

celestial sphere   An imaginary sphere surrounding Earth into which the stars are imagined to be fixed. For hundreds of years, people believed such a sphere (or bowl) really existed. Today, however, astronomers use the concept as a convenient fiction to describe the position of stars relative to one another.

chain reaction   *See* nuclear fission.

circumpolar stars   Stars near the celestial North Pole; from many locations on Earth, these stars never set.

closed universe   A universe that is finite and without boundaries. A universe with density above the critical value is necessarily closed (*see* critical density). An open universe, in contrast, will expand forever because its density is insufficient to halt the expansion.

comet   Also thought of as a "dirty snowball." A small celestial body composed mainly of ice and dust that completes highly eccentric orbits around the Sun. As a comet approaches the Sun, some of its material is vaporized and ionized to create a gaseous head (coma) and two long tails (one made of dust, one of ions).

conjunction   The apparent coming together of two celestial objects in the sky.

constellations   Arbitrary formations of stars that are perceived as figures or designs. There are 88 official constellations in the Northern and Southern Hemispheres. (*See* Appendix C.)

convective motion   A gas-flow pattern created by the rising movement of warm gases (or liquids) and the sinking movement of cooler gases (or liquids).

core   The innermost region of a planet or star.

core-collapse supernova   The extraordinarily energetic explosion that results when the core of a high-mass star collapses under its own gravity.

core-halo galaxy   *See* radio galaxy.

core hydrogen burning   The principal nuclear fusion reaction process of a star. The hydrogen at the star's core is fused into helium, and the small amount of mass lost is used to produce enormous amounts of energy.

corona   In astronomy, a corona might be a luminous ring appearing to surround a celestial body, the luminous envelope of ionized gas outside the Sun's chromosphere, or a large upswelling in the mantle of the surface of a planet or moon that takes the form of concentric fissures and that is an effect of volcanic activity. *See also* calderas.

cosmic microwave background (CMB)   The highly redshifted photons left behind by the Big Bang and detectable today throughout all space as radio-wavelength radiation, indicating a black-body temperature for the universe of 2.73 K.

cosmological principle   A cornerstone assumption about the nature of the universe. It holds that the universe exhibits two key properties: homogeneity (sameness of structure on the largest scale) and isotropy (appears the same in all directions).

cosmological redshift   The lengthening of the wavelengths of electromagnetic radiation caused by the expansion of the universe.

cosmology   The study of the origin, structure, and evolution of the universe.

crater   The Latin word for "bowl," it refers to the shape of depressions in the Moon or other celestial objects created (mostly) by meteoroid impacts.

critical density   The density of matter in the universe that represents the division between a universe that expands infinitely (unbound, or open) and one that will ultimately collapse (bound, or closed). The density of the universe determines whether it will expand forever or end with a conflagration as dramatic as the Big Bang.

crust   The surface layer of a planet.

dark energy   A catch-all phrase used to describe an energy of unknown origin that drives the apparent acceleration of the expansion of the universe.

dark halo   The region surrounding the Milky Way and other galaxies that contains dark matter. The shape of the dark halo can be probed by examining in detail the effects its mass has on the rotation of the galaxy.

dark matter   A catch-all phrase used to describe an apparently abundant substance in the universe of unknown composition. Dark matter is 100 times more abundant than luminous matter on the largest scales.

differential rotation   A property of anything that rotates and is not rigid. A spinning CD is a rigid rotator. A spinning piece of gelatin is not as rigid, and a spinning cloud of gas is even less so. For example, the atmospheres of the outer planets and of the Sun have equatorial regions that rotate at a different rate from the polar regions.

Drake Equation   Proposes a number of terms that help us make a rough estimate of the number of civilizations in our galaxy, the Milky Way.

dust lanes   Dark areas sometimes visible within emission nebulae and galaxies; the term most frequently refers to interstellar absorption apparent in edge-on spiral galaxies.

eccentric   An ellipse (or elliptical orbit) is called eccentric when it is noncircular. An ellipse with an eccentricity of 0 is a circle, and an ellipse with an eccentricity of close to 1 would be very oblong.

eclipse   An astronomical event in which one body passes in front of another so that the light from the occulted (shadowed) body is blocked. When the Sun, Moon, and Earth align, the Moon blocks the light of the Sun, resulting in a solar eclipse.

eclipsing binaries   *See* visual binaries.

ecliptic   The ecliptic traces the apparent path of the Sun against the background stars of the celestial sphere. This great circle is inclined at 23½ degrees relative to the celestial equator, which is the projection of the Earth's equator onto the celestial sphere.

electromagnetic radiation   Energy in the form of rapidly fluctuating electric and magnetic fields and including visible light in addition to radio, infrared, ultraviolet, x-ray, and gamma-ray radiation. Often arises from moving charges in atoms and molecules, though high-energy radiation can arise in other processes.

electromagnetic spectrum   The complete range of electromagnetic radiation, from radio waves to gamma waves and everything in between.

ellipse   A "flattened" circle drawn around two foci instead of a single center point.

elliptical galaxy   A galaxy with no discernible disk or bulge and that looks like an oval or circle of stars on the sky. The true shapes of ellipticals vary from elongated ("footballs") to spherical ("baseballs") to flattened ("hamburger buns"). Elliptical galaxies consist of old stars and appear to have little or no gas in them.

emission lines   Narrow regions of the spectrum where a particular substance is observed to emit its energy. These lines result from basic processes occurring on the smallest scales in an atom (such as electrons moving between energy levels).

emission nebulae   Glowing clouds of hot, ionized interstellar gas located near a young, massive star. (Singular, *emission nebula*.)

ephemeredes   Special almanacs that give the daily positions of various celestial objects for periods of several years.

escape velocity   The velocity necessary for an object to escape the gravitational pull of another object.

event horizon   Coincides with the Schwarzschild radius and is an imaginary boundary surrounding a concentration of mass, such as a collapsing star or black hole. Within the event horizon, no information of the events occurring there can be communicated to the outside, as at this distance not even the speed of light provides sufficient velocity to escape.

extrasolar planet   A planet orbiting a star other than the Sun. Over 100 such planets have been discovered, though mostly in highly elliptical orbits.

flat universe   The universe that results if its density is precisely at the critical level. It is flat in the sense that its space is defined by the rules of ordinary Euclidean geometry—parallel lines never cross.

focal length   The distance from a mirror surface to the point where parallel rays of light are focused.

focus   The point at which a mirror concentrates parallel rays of light that strike its surface.

frequency   The number of wave crests that pass a given point per unit of time. By convention, this is measured in hertz (equivalent to one crest-to-crest cycle per second, named in honor of the nineteenth-century German physicist Heinrich Rudolf Hertz and abbreviated Hz).

galactic bulge   Also called nuclear bulge. A swelling at the center of spiral galaxies. Bulges consist of old stars and extend out a few thousand light-years from the galactic centers.

galactic disk   The thinnest part of a spiral galaxy. The disk surrounds the nuclear bulge and contains a mixture of old and young stars, gas, and dust. In the case of the Milky Way, it extends out some 50,000 light-years from the Galactic center but is only about 1,000 light-years thick. It is the dust in the disk (a few hundred light-years thick) that creates the dark ribbon that runs the length of the Milky Way and limits the view of our own Galaxy (*see* dust lanes).

Galactic halo   A large (50,000 light-year radius) sphere of old stars surrounding the galaxy.

galactic nucleus   The core of a spiral galaxy. In the case of Seyfert (active) galaxies, the nucleus is extremely luminous in the radio wavelengths.

galaxy cluster   A gravitationally bound group of galaxies.

geocentric   Earth-centered; the geocentric model of the universe (or solar system) is one in which Earth is believed to be at the center of the universe (or solar system).

giant molecular clouds (GMCs)   Huge collections of cold (10 K to 100 K) gas that contain many millions of solar masses of molecular hydrogen. These clouds also contain other molecules (such as carbon dioxide) that can be imaged with radio telescopes. The cores of these clouds are often the sites of the most recent star formation.

gibbous   A word from Middle English that means "bulging"—an apt description of the Moon's shape between its first and third quarter phases.

globular clusters   Collections of a few hundred thousand stars, held together by their mutual gravitational attraction, that are found in highly eccentric orbits above and below the galactic disk.

gnomon   Any object designed to project a shadow used as an indicator. The upright part of a sundial is a gnomon.

H-R diagram   Short for Hertzsprung-Russell diagram. A graphical plot of luminosity versus temperature for a group of stars. Used to determine the age of clusters of stars.

heliocentric   Sun-centered; describes our solar system, in which the planets and other bodies orbit the Sun.

helium flash   A stellar explosion produced by the rapid increase in the temperature of a red giant's core driven by the fusion of helium.

homogeneity   *See* cosmological principle.

Hubble's Law   The linear relationship between the velocity of a galaxy's recession and the galaxy's distance from us. Simply stated, the law says that the recessional velocity is directly proportional to the distance. Used to determine the age of the universe.

inflationary epoch   A time soon after the Big Bang when the universe was puffed up suddenly, increasing in size by a factor of 10^{50} in an instant. This inflation could account for the incredible sameness, or uniformity, of the universe, even in regions that (without inflation) could never have been in contact with one another.

interferometer   A combination of telescopes linked together to create the equivalent (in terms of resolution) of a giant telescope. This computing-intensive method greatly increases resolving power.

interstellar matter   The material found between stars. Refers to the gas and dust thinly distributed throughout space, the matter from which the stars are formed. About 5 percent of our Galaxy's mass is contained in its gas and dust. The remaining 95 percent is in stars.

interstellar medium   *See* interstellar matter.

intrinsic variable   *See* variable star.

irregular galaxy   A galaxy type lacking obvious structure but containing lots of raw materials for the creation of new stars, and often, many young, hot stars.

isotropy   *See* cosmological principle.

jovian planets   The gaseous planets in our solar system farthest from the Sun (with the exception of Pluto, which is classed as neither terrestrial nor jovian): Jupiter, Saturn, Uranus, and Neptune.

Kelvin scale   The Kelvin (K) temperature scale is tied to the Celsius (C) temperature scale and is useful because there are no negative Kelvin temperatures. Absolute zero (0 K) is the coldest temperature that matter can attain. At this temperature, the atoms in matter would stop jiggling around all together. 0 K corresponds to approximately –273 C (-459.4 F).

leading face and **trailing face**   Moons that are tidally locked to their parent planet have a leading face and a trailing face; the leading face always faces in the direction of the orbit, the trailing face away from it. Many moons of the outer gas planets are in such locked orbits.

libration   The slow oscillation of the Moon (or other satellite, natural or artificial) as it orbits a larger celestial body. Lunar libration gives us glimpses of a very small portion of the far side of our Moon.

light pollution   The effect of poorly planned lighting fixtures (such as street and building lighting) that allow light to be directed upward into the sky. Light pollution washes out the contrast between the night sky and stars.

light-year   The distance light travels in one year: approximately 5.88 trillion miles (9.46 trillion km). For interstellar measurements, astronomers use the light-year as a basic unit of distance.

(The) Local Group   A galaxy cluster; the gravitationally bound group of galaxies that includes the Milky Way, Andromeda, and other smaller galaxies.

luminosity   The total energy radiated by a star each second. Luminosity is a quality intrinsic to the star; magnitude might or might not be.

magnetosphere   A zone of electrically charged particles trapped by a planet's magnetic field. The magnetosphere lies far above the planet's atmosphere.

magnitude   A system for classifying stars according to apparent brightness (*see* luminosity). The human eye can detect stars with magnitudes from 1 (the brightest) to 6 (the faintest). A first magnitude star is 100 times brighter than a sixth magnitude star. Absolute magnitude is another name for luminosity, but apparent magnitude is the amount of energy emitted by a star and striking some surface or detection device (including our eyes). Apparent magnitude varies with distance.

main sequence   When the temperature and luminosity of a large number of stars are plotted, the points tend to fall mostly in a diagonal region across the plot. The main sequence is this well-defined region of the Hertzsprung-Russell diagram (H-R diagram) in which stars spend most of their lifetime. The Sun will be a main sequence star for about 10 billion years.

mantle   The layer of a planet beneath its crust and surrounding its core.

maria (pronounced *MAH-ree-uh*)   The plural of "mare" (pronounced *MAR-ay*, with a long *a* at the end), this word is Latin for "seas." Maria are dark-grayish plains on the lunar surface that resembled bodies of water to early observers.

meteor   The term for a bright streak across the night sky—a "shooting star." A *meteoroid* is the object itself, a rocky object that is typically a tiny fragment lost from a comet or an asteroid. A *micrometeoroid* is a very small meteoroid. The few meteoroids that are not consumed in Earth's atmosphere reach the ground as *meteorites*.

meteor shower   When Earth's orbit intersects the debris that litters the path of a comet, we see a meteor shower. These happen at regular times during the year.

meteorite   *See* meteor.

meteoroid   *See* meteor.

micrometeoroid   *See* meteor.

millisecond pulsar   A neutron star rotating at some 1,000 revolutions per second and emitting energy in extremely rapid pulses.

minor planet   *See* asteroid.

nebula   A term with several applications in astronomy but used most generally to describe any fuzzy patch seen in the sky. Nebulae (the plural form of the term) are often (though not always) vast clouds of dust and gas.

neutron star   The super-dense compact remnant of a massive star, one possible survivor of a supernova explosion. It is supported by degenerate neutron pressure, not fusion. It is an entire star with the density of an atomic nucleus.

nova   A star that suddenly and very dramatically brightens, resulting from the triggering of nuclear fusion caused by the accretion of material from a binary companion star.

nuclear fission   A nuclear reaction in which an atomic unit splits into fragments, thereby releasing energy. In a fission reactor, the split-off fragments collide with other nuclei, causing them to fragment, until a chain reaction is under way.

nuclear fusion   A nuclear reaction that produces energy by joining atomic nuclei. Although the mass of a nucleus produced by joining two nuclei is less than that of the sum of the original two nuclei, the mass is not lost; rather, it is converted into large amounts of energy.

objective lens   In a telescope, the lens that first receives light from the observed object and forms an image.

open universe   A universe whose density is below the critical value (*see* critical density). In contrast to a closed universe, an open universe will expand forever because its density is insufficient to halt the expansion.

optical window   An atmospheric property that allows visible light to reach us from space.

orbital period   The time required for an object to complete one full orbit around another object. The orbital period of the Earth around the Sun, for example, is a fraction over 365 days.

planetary nebula   The ejected gaseous envelope of a red giant star. This shell of gas is lit up by the ultraviolet photons that escape from the hot, white dwarf star that remains. (This term is sometimes confusing because it has nothing to do with planets.)

planetary transit   An event in which a planet passes in front of the disk of the Sun (or the disk of another star in an extrasolar planetary system). These events in distant systems can be used to probe the properties of the transiting planet.

planetesimals   Embryonic planets in an early formative stage. Planetesimals, which are usually the size of small moons, develop into protoplanets, immature but full-scale planets. It is the protoplanets that go on to develop into mature planets as they cool.

precession   The slow change in the direction of the axis of a spinning object (such as the Earth), caused by an external influence or influences (such as the gravitational fields of the Sun and the Moon).

primary mirror   In a reflecting telescope, the mirror that first reflects light from the observed object.

primordial synthesis   The fusion reactions that occurred in the early universe at temperatures of about 10^{10} K. These fusion reactions produced helium and a small amount of lithium.

proper motion   Motion of a star determined by measuring the angular displacement of a target star relative to more distant background stars. Measurements are taken over long periods of time, and the result is an angular velocity (measured, for example, in arcseconds/year). If the distance to the star is known, this angular displacement can be converted into a transverse velocity in km/s (*see* transverse component).

protoplanet   *See* planetesimals.

pulsar   A rapidly rotating neutron star whose magnetic field is oriented such that it sweeps across Earth with a regular period.

pulsating variable   *See* variable star.

quasar   Short for "quasi-stellar radio source." Quasars are bright, distant, tiny objects that produce the luminosity of 100 to 1,000 galaxies within a region the size of a solar system.

radar   Short for "radio detection and ranging." In astronomy, radio signals are sometimes used to measure the distance of planets and other objects in the solar system.

radial component   *See* transverse component.

radio galaxy   A member of an active galaxy subclass of elliptical galaxies. Radio galaxies are characterized by strong radio emissions and, in some cases, narrow jets and wispy lobes of emissions located hundreds of thousands of light-years from the nucleus.

radio jets   Narrow beams of ionized material that have been ejected at relativistic velocities from a galaxy's nucleus.

radio lobe   The diffuse or wispy radio emissions found at the end of a radio jet. In some radio galaxies, the lobe emission dominates; in others, the jet emission dominates.

radio telescope   An instrument, usually a very large dish-type antenna connected to a receiver and recording and/or imaging equipment, used to observe radio-wavelength electromagnetic radiation emitted by stars and other celestial objects.

radio window   A property of Earth's atmosphere that allows some radio waves from space to reach Earth and that allows some radio waves broadcast from Earth to penetrate the atmosphere.

radioactive decay   The natural process whereby a specific atom or isotope is converted into another specific atom or isotope at a constant and known rate. By measuring the relative abundance of parent-and-daughter nuclei in a given sample of material (such as a meteorite), it is possible to determine the age of the sample.

red giant   A late stage in the career of stars about as massive as the Sun. More massive stars in their giant phase are referred to as supergiants. The relatively low surface temperature of this stage produces its red color.

redshift   An increase in the detected wavelength of electromagnetic radiation emitted by a celestial object as the recessional velocity between it and the observer increases. The name derives from the fact that lengthening the wavelength of visible light tends to redden the light that is observed. By analogy, "redshift" is applied to the lengthening of any electromagnetic wave.

refracting telescope   Also called a refractor. A telescope that creates its image by refracting (bending) light rays with lenses.

resolving power   The ability of a telescope (optical or radio telescope) to render distinct, individual images of objects that are close together.

retrograde   An orbit that is backward or contrary to the orbital direction of the other planets.

Schwarzschild radius   The radius of an object with a given mass at which the escape velocity equals the speed of light. As a rule of thumb, the Schwarzschild radius of a black hole (in km) is approximately three times its mass in solar masses, so a 5-solar mass black hole has a Schwarzschild radius of about $5 \times 3 = 15$ km.

seeing   The degradation of optical telescopic images as a result of atmospheric turbulence. "Good seeing" denotes conditions relatively free from such atmospheric interference.

Seyfert galaxy   A type of active galaxy, resembling a spiral galaxy, whose strong radio-wavelength emissions and emission lines come from a small region at its core.

sidereal day   A day measured from star rise to star rise. The sidereal day is 3.9 minutes shorter than the solar day.

sidereal month   The period of 27.3 days that it takes the Moon to orbit once around Earth. *See also* synodic month.

sidereal year   The time it takes Earth to complete one circuit around the Sun with respect to the stars.

singularity   The infinitely dense remnant of a massive core collapse.

solar day   A day measured from sunup to sunup (or noon to noon, or sunset to sunset). Slightly longer than a sidereal day. *See* sidereal day.

solar flares   Explosive events that occur in or near an active region on the Sun's surface.

solar nebula   The vast primordial cloud of gas and dust from which (it has been theorized) the Sun and solar system were formed.

solar wind   A continuous stream of radiation and matter that escapes from the Sun. Its effects can be seen in how it blows the tails of a comet approaching the Sun.

spectral lines   A system for classifying stars according to their surface temperature as measured by their spectra. The presence or absence of certain spectral lines is used to place stars in a spectral class. *See also* spectroscope.

spectrometer   *See* spectroscope.

spectroscope   An instrument that passes incoming light through a slit and prism, splitting it into its component colors. A spectrometer is an instrument capable of precisely measuring the spectrum thus produced. Substances produce characteristic spectral lines or emission lines, which act as the "fingerprint" of the substance, enabling identification of it.

spectroscopic binaries   *See* visual binaries.

spicules   Jets of matter expelled from the Sun's photosphere region into the chromosphere above it.

spiral arms   Structures found in spiral galaxies apparently caused by the action of spiral density waves.

spiral density waves   Waves of compression that move around the disk of a spiral galaxy. These waves are thought to trigger clouds of gas into collapse, thus forming hot, young stars. It is mostly the ionized gas around these young, massive stars that we observe as spiral arms.

spiral galaxy   A galaxy characterized by a distinctive structure consisting of a thin disk surrounding a galactic bulge. The disk is dominated by bright, curved arcs of emissions known as spiral arms. The rotation curves of spiral galaxies indicate the presence of large amounts of dark matter.

standard candle   Any object whose luminosity is well known. Its measured brightness can then be used to determine how far away the object is. The brightest standard candles can be seen from the greatest distances.

standard solar model   Our current picture of the structure of the Sun. The model seeks to explain the observable properties of the Sun and also to describe properties of its mostly unobservable interior.

stellar occultation   An astronomical event that occurs when a planet passes in front of a star, dimming the star's light (as seen from Earth). The exact way in which the light dims can reveal details, for example, in the planet's atmosphere.

summer solstice   On or about June 21; the longest day in the Northern Hemisphere, marking the beginning of summer.

sunspots   Irregularly shaped dark areas on the face of the Sun. They appear dark because they are cooler than the surrounding material. They are tied to the presence of magnetic fields at the Sun's surface.

supercluster   A group of galaxy clusters. The Local Supercluster contains some 10^{15} solar masses.

superluminal motion   A term for the apparent "faster than light" motion of blobs of material in some radio jets. This effect results from radio-emitting blobs moving at high velocity toward the observer.

supernova   The explosion accompanying the death of a massive star as its core collapses.

synchronous orbit   A celestial object is in synchronous orbit when its period of rotation is equal to its average orbital period; the Moon, in synchronous orbit, presents only one face to Earth.

synchrotron radiation   Synchrotron radiation arises when charged particles (e.g., electrons) are accelerated by strong magnetic fields. Some of the emissions from radio galaxies is synchrotron.

synodic month   The period of 29.5 days that the Moon requires to cycle through its phases, from new moon to new moon. *See also* sidereal month.

telescope   A word from Greek roots meaning "far-seeing." Optical telescopes are arrangements of lenses and/or mirrors designed to gather visible light efficiently enough to enhance resolution and sensitivity. *See also* radio telescope.

terminator   The boundary separating light from dark, the daytime from nighttime hemispheres of the Moon (or other planetary or lunar bodies).

terrestrial planets   The planets in our solar system closest to the Sun: Mercury, Venus, Earth, and Mars.

thought experiment   A systematic hypothetical or imaginary simulation of reality, used as an alternative to actual experimentation when such experimentation is impractical or impossible.

tidal bulge   The deformation of one celestial body caused by the gravitational force of another extended celestial body. The Moon creates an elongation of Earth's oceans—a tidal bulge.

time dilation   The apparent slowing of time (as perceived by an outside observer) as an object approaches the event horizon of a black hole or moves at very high velocity.

trailing face   *See* leading face.

transit   When an object crosses the imaginary half-circle on the sky that runs from north to south (the meridian), it is said to transit.

transverse component   Stellar movement across the sky, perpendicular to our line of sight. The radial component is motion toward or away from us. True space motion is calculated by combining the observed transverse and radial components.

triangulation   An indirect method of measuring distance derived by geometry or trigonometry using a known baseline and two angles from the baseline to the object.

tropical year   A year measured from equinox to equinox. *See also* sidereal year.

universal recession   The apparent general movement of all galaxies away from us first observed by Edwin Hubble. This observation does not mean we are at the center of the expansion. Any observer located anywhere in the universe should see the same redshift.

Van Allen belts   Named for their discoverer, American physicist James A. Van Allen, these are vast doughnut-shaped zones of highly energetic, charged particles that are trapped in the magnetic field of Earth. The zones were discovered in 1958.

variable star   A star that periodically changes in brightness. A cataclysmic variable is a star, such as a nova or supernova, that changes in brightness suddenly and dramatically as a result of interaction with a binary companion star, while an intrinsic variable changes brightness because of rapid changes in its diameter. Pulsating variables are intrinsic variables that vary in brightness in a fixed period or span of time and are useful distance indicators.

vernal equinox   On this date (usually March 21), day and night are of equal duration because the Sun's apparent course intersects the celestial equator at these times.

visual binaries   Binary stars that can be resolved from Earth. Spectroscopic binaries are too distant to be seen as distinct points of light, but they can be observed with a spectroscope. In this case, the presence of a binary system is detected by noting Doppler-shifting spectral lines as the stars orbit one another. If the orbit of one star in a binary system periodically eclipses its partner, it's possible to monitor the variations of light emitted from the system and thereby gather information about orbital motion, mass, and radii. These binaries are called eclipsing binaries.

water hole   The span of the radio spectrum from 18 cm to 21 cm, which many researchers believe is the most likely wavelength on which extraterrestrial broadcasts will be made. The name is a little astronomical joke—the hydrogen (H) and hydroxyl (OH) lines are both located in a quiet region of the radio spectrum, a region where there isn't a lot of background noise. Because H and OH add up to H_2O (water), this dip in the spectrum is called the water hole.

wavelength   The distance between two adjacent wave crests (high points) or troughs (low points). By convention, this distance is measured in meters or decimal fractions thereof.

white dwarf   The remnant core of a red giant after it has lost its outer layers as a planetary nebula. Because fusion has halted, the carbon-oxygen core is supported against further collapse only by the pressure supplied by densely packed electrons.

winter solstice   On or about December 21; in the Northern Hemisphere, the shortest day and the start of winter.

zonal flow   The prevailing east-west wind pattern that is found on Jupiter.

Upcoming Eclipses

Total Solar Eclipses

Date	Duration of Totality (in Minutes)	Where Visible
2005 Apr. 8	0.7	South Pacific Ocean
2006 Mar. 29	4.1	Africa, Asia Minor, Russia, the C.I.S.
2008 Aug. 1	2.4	Arctic Ocean, Siberia, China
2009 July 22	6.6	India, China, South Pacific
2010 July 11	5.3	South Pacific
2012 Nov. 13	4.0	Northern Australia, South Pacific
2013 Nov. 3	1.7	Atlantic Ocean, Central Africa
2015 Mar. 20	4.1	North Atlantic, Arctic Ocean
2016 Mar. 9	4.5	Indonesia, Pacific Ocean
2017 Aug. 21	2.0	Pacific Ocean, United States, Atlantic Ocean

Total Lunar Eclipses

2004 May 4
2004 Oct. 28
2007 Mar. 3
2007 Aug. 28
2008 Feb. 21

The Constellations

The constellations visible in the night sky change throughout the year. The four charts that follow are seasonal views (March, June, September and December) of the night sky. While the charts represent the sky as seen from altitude of 34 degrees north, they are useful anywhere in the continental United States. The directions (North, South, East, West) are indicated on each chart.

Note that the North Star (Polaris) and the constellations near it—the circumpolar constellations—are visible throughout the year. These include Cassiopeia, Ursa Major (the "Big Dipper"), Draco and others. The gray band in each chart represents the plane of the Milky Way, our home galaxy, which is visible only in the very darkest skies.

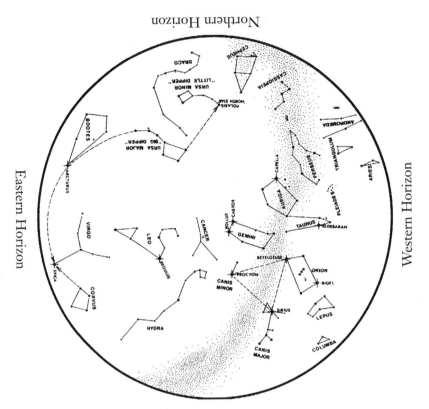

The Night Sky in March

Latitude of chart is 34°N, but it is practical throughout the continental United States.

To use: Hold chart vertically and turn it so the direction you are facing shows at the bottom.

Chart time (Local Standard):

10 P.M. First of month

9 P.M. Middle of month

8 P.M. Last of month

Star Chart from Griffith Observer, *Griffith Obervatory, Los Angeles*

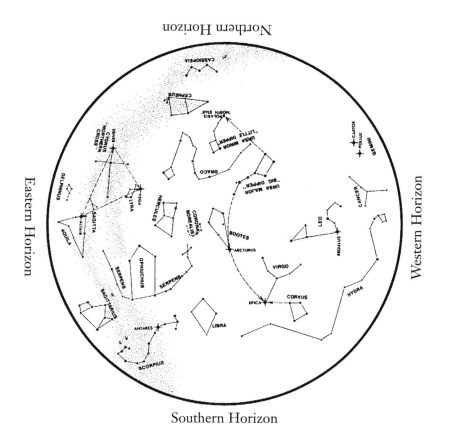

The Night Sky in June

Latitude of chart is 34°N, but it is practical throughout the continental United States.

To use: Hold chart vertically and turn it so the direction you are facing shows at the bottom.

Chart time (Local Standard):

10 P.M. First of month

9 P.M. Middle of month

8 P.M. Last of month

Star Chart from Griffith Observer, *Griffith Observatory, Los Angeles*

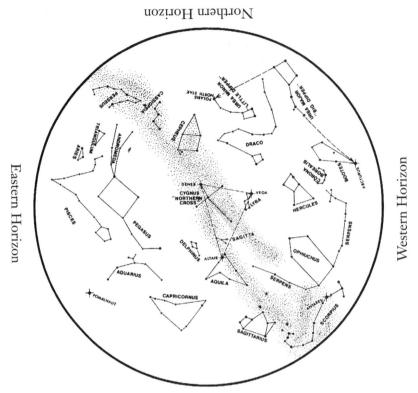

Southern Horizon

The Night Sky in September

Latitude of chart is 34°N, but it is practical throughout the continental United States.

To use: Hold chart vertically and turn it so the direction you are facing shows at the bottom.

Chart time (Local Standard):

10 P.M. First of month

9 P.M. Middle of month

8 P.M. Last of month

Star Chart from Griffith Observer, *Griffith Observatory, Los Angeles*

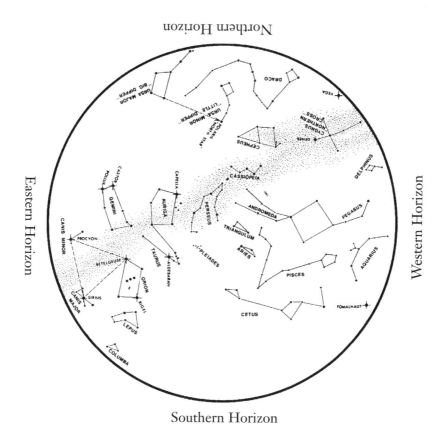

Southern Horizon

The Night Sky in December

Latitude of chart is 34°N, but it is practical throughout the continental United States.

To use: Hold chart vertically and turn it so the direction you are facing shows at the bottom.

Chart time (Local Standard):

10 P.M. First of month

9 P.M. Middle of month

8 P.M. Last of month

Star Chart from Griffith Observer, *Griffith Observatory, Los Angeles*

The Messier Catalog

French astronomer Charles Messier (1730–1817) was the first person to compile a systematic catalog of nebulae and star clusters. His purpose in this was to make comet hunting easier by taking careful note of permanent deep-sky objects that might be mistaken for comets. In Messier's time, the term *nebula* was applied to any blurry celestial object. The Messier Catalog that resulted is valued today by amateur astronomers as a checklist of deep sky objects easily observable with even a modest telescope. The 110 Messier objects are considered the most interesting deep-sky objects visible in the Northern Hemisphere. For images of all the Messier objects, see www.seds.org/messier.

Legend

M# = Messier Catalog number

NGC# = *New General Catalog Number* (The *NGC* is a modern work, which, among professional astronomers, has superseded the Messier Catalog.)

Con = The constellation in which the object is found. See Appendix C for the full names of the constellations abbreviated here.

RA = Right ascension in hours and minutes. See Chapter 1 for an explanation.

DEC = Declination in degrees and minutes. See Chapter 1 for an explanation.

vMag = Visual Magnitude of the object. See Chapter 13 for an explanation of magnitude.

Type = Type of object. The abbreviations are as follows:

DS	Double star
E (with a number)	Elliptical galaxy
EN	Emission nebula (glowing gas)
ERN	Emission and reflection nebula
GC	Globular cluster (stars)
Irr	Irregular galaxy
OC	Open cluster (stars)
PN	Planetary nebula
S	Spiral galaxy
SB	Barred-spiral galaxy
SCL	Star cloud
SNR	Supernova remnant

Dia" = Angular diameter (where known) of the object. This number is in seconds of arc ($\frac{1}{3,600}$ degree). See Chapter 1 for an explanation of angular measurement.

Name = The common name of the object.

M#	NGC#	Con	RA	Dec	vMag	Type	Dia"	Name
1	1952	Tau	05h 34.5m	+22d 01'	8.4	SNR	300	Crab Nebula
2	7089	Aqr	21h 33.5m	−00d 49'	6.5	GC		
3	5272	CVn	13h 42.2m	+28d 23'	6.4	GC		
4	6121	Sco	16h 23.6m	−26d 32'	5.9	GC		
5	5904	Ser	15h 18.6m	+02d 05'	5.8	GC		
6	6405	Sco	17h 40.1m	−32d 13'	4.2	OC		Butterfly Cluster
7	6475	Sco	17h 53.9m	−34d 49'	3.3	OC		
8	6523	Sgr	18h 03.8m	−24d 23'	5.8	EN	2400	Lagoon Nebula
9	6333	Oph	17h 19.2m	−18d 31'	7.9	GC		
10	6254	Oph	16h 57.1m	−04d 06'	6.6	GC		
11	6705	Sct	18h 51.1m	−06d 16'	5.8	OC	750	Wild Duck Cluster
12	6218	Oph	16h 47.2m	−01d 57'	6.6	GC		
13	6205	Her	16h 41.7m	+36d 28'	5.9	GC		Hercules Cluster
14	6402	Oph	17h 37.6m	−03d 15'	7.6	GC		
15	7078	Peg	21h 30.0m	+12d 10'	6.4	GC		
16	6611	Ser	18h 18.8m	−13d 47'	6.0	EN		Eagle Nebula
17	6618	Sgr	18h 20.8m	−16d 11'	7.0	EN	1200	Swan Nebula (also known as Omega Nebula)
18	6613	Sgr	18h 19.9m	−17d 08'	6.9	OC		
19	6273	Oph	17h 02.6m	−26d 16'	7.2	GC		
20	6514	Sgr	18h 02.6m	−23d 02'	8.5	ERN	900	Trifid Nebula
21	6531	Sgr	18h 04.6m	−22d 30'	5.9	OC		
22	6656	Sgr	18h 36.4m	−23d 54'	5.1	GC		
23	6494	Sgr	17h 56.8m	−19d 01'	5.5	OC		
24	6603	Sgr	18h 16.9m	−18d 29'	4.5	SCL		(star cloud)
25	4725	Sgr	18h 31.6m	−19d 15'	4.6	OC		
26	6694	Sct	18h 45.2m	−09d 24'	8.0	OC		
27	6853	Vul	19h 59.6m	+22d 43'	8.1	PN	420	Dumbbell Nebula
28	6626	Sgr	18h 24.5m	−24d 52'	6.9	GC		
29	6913	Cyg	20h 23.9m	+38d 32'	6.6	OC		
30	7099	Cap	21h 40.4m	−23d 11'	7.5	GC		

M#	NGC#	Con	RA	Dec	vMag	Type	Dia"	Name
31	224	And	00h 42.7m	+41d 16'	3.4	SB	14400	Great Andromeda Spiral
32	221	And	00h 42.7m	+40d 52'	8.2	E2		(M31 companion)
33	598	Tri	01h 33.9m	+30d 39'	5.7	S	3600	Triangulum Spiral
34	1039	Per	02h 42.0m	+42d 47'	5.2	OC		
35	2168	Gem	06h 08.9m	+24d 20'	5.1	OC		
36	1960	Aur	05h 36.1m	+34d 08'	6.0	OC		
37	2099	Aur	05h 52.4m	+32d 33'	5.6	OC		
38	1912	Aur	05h 28.7m	+35d 50'	6.4	OC		
39	7092	Cyg	21h 32.2m	+48d 26'	4.6	OC		
40	WIN4	UMa	12h 22.4m	+58d 05'	8.0	DS		Winnecke 4
41	2287	CMa	06h 47.0m	−20d 44'	4.5	OC		
42	1976	Ori	05h 35.4m	−05d 27'	4.0	EN		Great Orion Nebula
43	1982	Ori	05h 35.6m	−05d 16'	9.0	EN		Orion Nebula (part)
44	2632	Cnc	08h 40.1m	+19d 59'	3.1	OC	5400	Beehive Cluster
45	1432	Tau	03h 47.0m	+24d 07'	1.2	OC	7200	Pleiades
46	2437	Pup	07h 41.8m	−14d 49'	6.1	OC		
47	2422	Pup	07h 36.6m	−14d 30'	4.4	OC		
48	2548	Hya	08h 13.8m	−05d 48'	5.8	OC		
49	4472	Vir	12h 29.8m	+08d 00'	8.4	E4		
50	2323	Mon	07h 03.2m	−08d 20'	5.9	OC		
51	1945	CVn	13h 29.9m	+47d 12'	8.1	S		Whirlpool galaxy
52	7654	Cas	23h 24.2m	+61d 35'	6.9	OC		
53	5024	Com	13h 12.9m	+18d 10'	7.7	GC		
54	6715	Sgr	18h 55.1m	−30d 29'	7.7	GC		
55	6809	Sgr	19h 40.0m	−30d 58'	7.0	GC		
56	6779	Lyr	19h 16.6m	+30d 11'	8.2	GC		
57	6720	Lyr	18h 53.6m	+33d 02'	9.0	PN		Ring Nebula
58	4579	Vir	12h 37.7m	+11d 49'	9.8	SB		
59	4621	Vir	12h 42.0m	+11d 39'	9.8	E3		
60	4649	Vir	12h 43.7m	+11d 33'	8.8	E1		
61	4303	Vir	12h 21.9m	+04d 28'	9.7	S		
62	6266	Oph	17h 01.2m	−30d 07'	6.6	GC		
63	5055	CVn	13h 15.8m	+42d 02'	8.6	SB		Sunflower galaxy

M#	NGC#	Con	RA	Dec	vMag	Type	Dia"	Name
64	4826	Com	12h 56.7m	+21d 41'	8.5	SB		Black Eye galaxy
65	3623	Leo	11h 18.9m	+13d 05'	9.3	SB		
66	3627	Leo	11h 20.2m	+12d 59'	9.0	SB		
67	2682	Cnc	08h 50.4m	+11d 49'	6.9	OC		
68	4590	Hya	12h 39.5m	−26d 45'	8.2	GC		
69	6637	Sgr	18h 31.4m	−32d 21'	7.7	GC		
70	6681	Sgr	18h 43.2m	−32d 18'	8.1	GC		
71	6838	Sge	19h 53.8m	+18d 47'	8.3	GC		
72	6981	Aqr	20h 53.5m	−12d 32'	9.4	GC		
73	6994	Aqr	20h 58.9m	−12d 38'	5.0	OC		
74	628	Psc	01h 36.7m	+15d 47'	9.2	S		
75	6864	Sgr	20h 06.1m	−21d 55'	8.6	GC		
76	650-1	Per	01h 42.4m	+51d 34'	11.5	PN		Little Dumbbell
77	1068	Cet	02h 42.7m	−00d 01'	8.8	SB		
78	2068	Ori	05h 46.7m	+00d 03'	8.0	PN		
79	1904	Lep	05h 24.5m	−24d 33'	8.0	GC		
80	6093	Sco	16h 17.0m	−22d 59'	7.2	GC		
81	3031	UMa	09h 55.6m	+69d 04'	6.8	SB		
82	3034	UMa	09h 55.8m	+69d 41'	8.4	Irr		"Exploding" galaxy
83	5236	Hya	13h 37.0m	−29d 52'	10.1	S		
84	4374	Vir	12h 25.1m	+12d 53'	9.3	E1		
85	4382	Com	12h 25.4m	+18d 11'	9.3	Ep		
86	4406	Vir	12h 26.2m	+12d 57'	9.2	E3		
87	4486	Vir	12h 30.8m	+12d 24'	8.6	E1		
88	4501	Com	12h 32.0m	+14d 25'	9.5	SB		
89	4552	Vir	12h 35.7m	+12d 33'	9.8	E0		
90	4569	Vir	12h 36.8m	+13d 10'	9.5	SB		
91	4548	Com	12h 35.4m	+14d 30'	10.2	SB		
92	6341	Her	17h 17.1m	+43d 08'	6.5	GC		
93	2447	Pup	07h 44.6m	−23d 52'	6.2	OC		
94	4736	CVn	12h 50.9m	+41d 07'	8.1	SB		
95	3351	Leo	10h 44.0m	+11d 42'	9.7	SB		
96	3368	Leo	10h 46.8m	+11d 49'	9.2	SB		

M#	NGC#	Con	RA	Dec	vMag	Type	Dia"	Name
97	3587	UMa	11h 14.8m	+55d 01'	11.2	PN		Owl Nebula
98	4192	Com	12h 13.8m	+14d 54'	10.1	SB		
99	4254	Com	12h 18.8m	+14d 25'	9.8	S		
100	4321	Com	12h 22.9m	+15d 49'	9.4	S		
101	5457	UMa	14h 03.2m	+54d 21'	7.7	S		Pinwheel galaxy
102	5866	Dra	15h 06.5m	+55d 46'	10.0	E6		
103	581	Cas	01h 33.2m	+60d 42'	7.4	OC		
104	4594	Vir	12h 40.0m	−11d 37'	8.3	SB		Sombrero galaxy
105	3379	Leo	10h 47.8m	+12d 35'	9.3	E1		
106	4258	CVn	12h 19.0m	+47d 18'	8.3	SB		
107	6171	Oph	16h 32.5m	−13d 03'	8.1	GC		
108	3556	UMa	11h 11.5m	+55d 40'	10.0	S		
109	3992	UMa	11h 57.6m	+53d 23'	9.8	SB		
110	205	And	00h 40.4m	+41d 41'	8.0	E6		(M31 companion)

Sources for Astronomers

Websites

A wealth of astronomy-related information and images are available on the World Wide Web. We've listed a few highlights for you to check out, but don't hesitate to use a good search engine (such as Google or Yahoo!) to look for additional or more specific information.

Institutions, Magazines, and Societies

Amateur Telescope Makers Association. Valuable for the do-it-yourselfer: **www.atmjournal.com.**

Astronomy magazine. Includes links to other sites: **www.astronomy.com/home.asp.**

Astrobiology.com. A site devoted to the maverick science of astrobiology: **www.astrobiology.com.**

Institute and Museum of History of Science, Florence, Italy. Includes a wealth of multimedia material (in English) devoted to the work of Galileo: **www.imss.fi.it.**

Jet Propulsion Laboratory, a NASA-related facility. A lot of information on solar system exploration projects. Many images. A must-see site: **www.jpl.nasa.gov.**

The Planetary Society. Founded by Carl Sagan to encourage the search for extraterrestrial life: **http://planetary.org.**

SETI Institute. Dedicated to the search for extraterrestrial civilizations: **www.seti-inst.edu.**

Sky & Telescope magazine. An important source of information (dates, directions) on astronomical events: **www.skypub.com.**

Some Observatories

Anglo-Australian Observatory: **www.aao.gov.au.**

Bradley Observatory (on-campus observatory at Agnes Scott College): **http://bradley.agnesscott.edu.**

Keck Observatory: **www2.keck.hawaii.edu.**

Mount Wilson Observatory: **www.mtwilson.edu.**

National Optical Astronomy Observatories: **www.noao.edu.**

National Radio Astronomy Observatory: **www.nrao.edu.**

Space Telescope Science Institute (source of *Hubble Space Telescope* images): **www.stsci.edu/public.html.**

Space Telescope Science Institute Press Release Page: **http://hubblesite.org/newscenter/.**

United States Naval Observatory (USNO). Information on timekeeping and sunrise and sunset times: **www.usno.navy.mil.**

Yerkes Observatory, University of Chicago: **http://astro.uchicago.edu/yerkes.**

Planetary Positions, Solar System

Planetary position calculator: **http://imagiware.com/astro/planets.cgi.**

Scale model solar system: **www.exploratorium.edu/ronh/solar_system.**

Guides to Events (Eclipses, Meteor Showers, and So On)

Abrams Planetarium Sky Calendar: **www.pa.msu.edu/abrams/SkyCalendar/Index.html.**

For solar eclipse information, including link to the SOHO site, with frequently updated solar images at many wavelengths: **http://umbra.nascom.nasa.gov/sdac.html**.

For comets and meteor showers: **http://comets.amsmeteors.org**.

Images and Catalogs

Astronomy Pictures of the Day: **http://antwrp.gsfc.nasa.gov/apod/astropix.html**.

The Messier Catalog (a visual catalog of Messier objects): **www.seds.org/messier**.

Sky View. Truly a virtual observatory: **http://skyview.gsfc.nasa.gov/**.

Links

The following sites serve as links to many other websites of interest to amateur and professional astronomers:

Astronomical Society of the Pacific. This site includes a link to *Mercury Magazine*: **www.astrosociety.org**.

American Astronomical Society (AAS). This site includes links to the online versions of the *Astronomical Journal* and the *Astrophysical Journal* in addition to information and statistics about the profession: **www.aas.org**.

Astronomical World Wide Web Resources: **http://cws.stsci.edu/astroweb/net-www**.

Other fascinating astronomy links: **www.open.hr/space/space/http.phtml**.

Magazines

Astronomy (Kalmbach Publishing, PO Box 1612, Waukesha, WI 53187). A popular and well-written monthly journal.

Griffith Observer (Griffith Observatory, 2800 E. Observatory Rd., Los Angeles, CA 90027). Concentrates on the history of astronomy.

Mercury Magazine (390 Ashton Avenue, San Francisco, CA 94112). A publication of the Astronomical Society of the Pacific (ASP); contains historical and scientific articles as well as excellent monthly columns.

Planetary Report (The Planetary Society, 65 N. Catalina Ave., Pasadena, CA 91106). Provides news on exploring the solar system and the search for extraterrestrial life.

Sky & Telescope (PO Box 9111, Belmont, MA 02178). Considered by some to be the standard for amateur astronomy magazines.

Books

As we told you, you're not alone in your fascination with astronomy. You'll find no shortage of books on the subject. The following are a few that are particularly useful.

Practical Guides

Berry, Richard. *Discover the Stars*. New York: Harmony Books, 1987.

Burnham, Robert, Jr. *Burnham's Celestial Handbook*. New York: Dover, 1978.

Carlson, Shawn, ed. *Amateur Astronomer*. New York: Wiley, 2000.

Charles, Jeffrey R. *Practical Astrophotography*. New York: Springer Verlag, 2000.

Covington, Michael A. *Astrophotography for the Amateur*. New York: Cambridge University Press, 1999.

Harrington, Philip S. *Star Ware: The Amateur Astronomer's Ultimate Guide to Choosing, Buying, and Using Telescopes and Accessories, 2nd ed.* New York: Wiley, 1998.

Levy, David H. *The Sky: A User's Guide*. Cambridge, England: Cambridge University Press, 1991.

Licher, David. *The Universe From Your Backyard*. Milwaukee: Kalmbach Publishing, 1988.

Mayall, R. Newton, et al. *The Sky Observer's Guide: A Handbook for Amateur Astronomers*. New York: Golden Books, 2000.

North, Gerald. *Advanced Amateur Astronomy*. New York: Cambridge University Press, 1997.

Tonkin, Stephen F., ed. *Amateur Telescope Making*. New York: Springer Verlag, 1999.

Webb, Stephen, *Where Is Everybody?* New York: Copernicus Books, 2002.

Guides to Events

Bishop, Roy, ed. *The Observer's Handbook.* Toronto: The Royal Astronomical Society of Canada, annual.

Westfall, John E., ed. *The ALPO Solar System Ephemeris.* San Francisco: Association of Lunar and Planetary Observers, annual.

Star and Lunar Atlases

Cook, Jeremy. *The Hatfield Photographic Lunar Atlas.* New York: Springer Verlag, 1999.

Dickinson, Terence, et al. *Mag 6 Star Atlas.* Barrington, NJ: Edmund Scientific, 1982.

Norton, Arthur P. *Norton's 2000.0, 18th ed.* Cambridge, MA: Sky Publishing Corporation, 1989.

Tirion, Wil. *Sky Atlas 2000.0.* Cambridge, MA: Sky Publishing Corporation and Cambridge University Press, 1981.

Introductory Textbooks and Popular Science Books

Chaisson, Eric, and Steve McMillan. *Astronomy: A Beginner's Guide to the Universe, 2nd ed.* Upper Saddle River, NJ: Prentice Hall, 1998.

Ferris, Timothy. *The Whole Shebang.* New York: Simon & Schuster, 1997.

Fraknoi, Andrew, et al. *Voyages Through the Universe.* Fort Worth, TX: Saunders College Publishing, 1997.

Hartman, William K. *Moons & Planets, 3d ed.* Belmont, CA: Wadsworth Publishers, 1993.

Kaufmann, William J., and Neil F. Comins. *Discovering the Universe, 4th ed.* New York: W. H. Freeman, 1996.

Life in the Universe

Davies, Paul. *The Fifth Miracle: The Search for the Origin of Life.* New York, NY: Simon & Schuster, 1999.

Goldsmith, Donald, and Tobias Owen. *The Search for Life in the Universe, 3rd ed.* Reading, MA: Addison-Wesley, 2001.

Grinspoon, David, *Lonely Planets: The Natural Philosophy of Alien Life*, New York: HarperCollins, 2003.

Index

X-Y-Z